AF615269

LOW-HORSEPOWER FUN AIRCRAFT YOU CAN BUILD

Other TAB Books by the author:

No. 2205 *Aircraft Metal Work*
No. 2230 *Restoration of Antique & Classic Planes*
No. 2254 *Man-Powered Aircraft*

No. 2267
$9.95

LOW-HORSEPOWER FUN AIRCRAFT YOU CAN BUILD

BY DON DWIGGINS

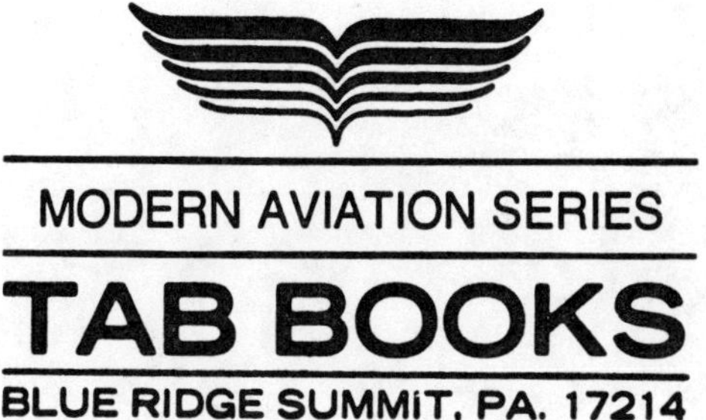

FIRST EDITION

FIRST PRINTING—DECEMBER 1979

Printed in the United States of America

Library of Congress Cataloging in Publication Data

Dwiggins, Don
Low-horsepower fun aircraft you can build

Includes index.
1. Home-airplanes build
I. Title
TL671. 2. D9 629. 133'34 79-22942
ISBN 0-8306-9710-1
ISBN 0-8306-2267-5 pbk.

Cover photos clockwise from top left: Bensen gyrocopter courtesy of Bensen Aircraft Corp.; Chevie Bird, VariEze, and VW powered Pazmany PL-4 prototype courtesy of Downie and Associates.

Contents

Chapter 1
In The Beginning

In the beginning there were Wilbur and Orville, plus an unknown number of other pioneer dreamers with sufficient mechanical skills to build their own aircraft, many of them outlandish, others amazingly well-designed, most of them sadly underpowered.

Since Leonardo da Vinci's time, soaring birds were the inspiration for winged heavier-than-air flying machines, and when inventors discovered that wing-flapping ornithopter designs were too complex, rigid-wing gliders were the choice.

Sir George Cayley in England, the Lilienthal brothers in Germany, Clement Ader in France, and a host of inventors in the United States learned by trial and error what would work and what wouldn't. Ultimately, it was the wing that sustained such craft, and what better way to design a wing than to make the airfoil curved, like a bird's?

The Wrights improved on the wing-curve tables of Lilienthal, and others like Hargreave in Australia, Montgomery in California, and Octave Chanute in the Midwest discovered the close relationship of center of lift and center of pressure to longitudinal stability.

Dihedral was copied from the upswept wings of the pigeon, but manned gliders, it was found, needed more than inherent stability for control—in fact, an unstable design was found better, with control surfaces added to guide the craft on its way.

Two schools of thought developed independently as to the best control system—the wing-warping method patented by Wilbur and Orville Wright, and the use of a midwing aileron, invented by the genius Alexander Graham Bell and patented in the name of a group known as the Aerial Experiment Association.

Had the Wrights searched the Patent Office files carefully, they would have been astounded to learn that a decade before their Kitty Hawk biplane first flew, a Frenchman, Louis Pierre Mouillard, had settled on the same concept in Cairo, Egypt, where he spent long hours watching the graceful flight of African vultures, and how they changed the shape of their wings in flight to maintain equilibrium when turning. Mouillard was right in his analysis of vulture flight, but he discovered something else that would work to the disadvantage of manned aircraft, using either wing-warping or ailerons for lateral control.

He wrote: "The birds effect a deformation of one wingtip so as to impede the air at that point and to turn upon it as a pivot." It was this "tip-feather drag" that would become "aileron" drag, or "adverse yaw", an annoying factor in airplanes that required use of a deflected vertical rudder to neutralize it when entering or leaving a turn.

Curiously, it also formed the basis for the successful design of the Gossamer Condor, Dr. Paul MacCready's amazing manpowered aircraft that won the coveted Kremer Award for successful MPA flight in 1977, just one example of how present-day designers of homebuilt aircraft are finding new ways to make their craft go where they want them to go.

Today we have Taras Kicuniuk's "dragerons"— differentially-controlled wingtip plates that act much like Mouillard's African vultures, by utilizing the tip-drag force to slew his Icarus hang glider around in a turn.

Many other novel ideas have been tried, some good, some bad. Where large aircraft corporations have the benefit of wind tunnels and extensive aeronautical engineering research facilities, the average homebuilder is dependent on limited resources and background information. Fortunately there is a number of tried and proven homebuilt designs to choose from, such as the Thorp T-18, the Bowers Fly Baby, the Rutan VariEze, and the Steen Skybolt, to name a few at random.

Why would anyone want to design and build an aircraft, when there are close to 200,000 active aircraft already flying, and thousands more coming off the production lines at Wichita, and elsewhere? Two basic motivations are involved—the inflationary cost of today's factory planes, and the desire of the builder to create an aircraft meeting his own personal requirements, both practically and aesthetically.

For starters, the homebuilder, or prospective homebuilder, should by all means join the Experimental Aircraft Association, the

First successful homebuilt was the Wright Flyer that started the Air Age at Kitty Hawk Dec. 17, 1903.

world's largest group of active builders and pilots of homebuilts, located at Box 229, Hales Corners, Wisconsin 53130. Over the last three decades the EAA has expanded from a small group of enthusiasts to a worldwide organization with more than 120,000 members in 91 countries and more than 500 active chapters.

EAA membership includes a subscription to Sport Aviation, a monthly periodical in which members share new ideas in design, construction, and operation of homebuilt aircraft. Some are far out, others so basic you wonder why you never thought of them yourself.

Each year the EAA stages a summer Fly-In at Wittman Field, Oshkosh, Wisconsin. This event has become, amazingly enough, the world's largest "airshow" with some 8,000 to 10,000 aircraft arriving for the week-long event, including perhaps 1,500 amateur-built planes, gyrocopters, powered hang gliders, etc. The Oshkosh Fly-In also attracts visitors numbering in the hundreds of thousands, an indicator of the interest today in "grass-roots" aviation.

Wittman Field is named for one of the real pioneers in the homebuilt movement—Steve Wittman, a noted racing pilot and designer of such craft as his Witt's Vee, a competition plane built to Formula One racing specifications, capable of flying up to three miles per minute. One of the old standbys in the homebuilt stable is another Wittman design, his eighth—the Wittman W-8 Tailwind, an amazing, two-place, side-by-side high-winger capable of speeds up to 170 mph using engines rated from 85 to 115 horsepower. Recently, Wittman installed a converted Oldsmobile V-8 aluminum block engine in his personal W-8.

Later on you'll read more about converted automotive engines, in both the low and high horsepower range, as builders scrounge the market for a more inexpensive way to power their homebuilts. Among them are the ubiquitous Volkswagen engines, Corvairs, Ford V-8s, and smaller engines from motorcycles, go-karts, and snowmobiles.

Each summer you'll find many static exhibits at the EAA's Oshkosh Fly-In, as well as well-attended forums at which experts in various fields of design, manufacture, and operation of aircraft keep the membership updated on new developments, and subjects of historical importance.

History, in fact, plays a big part in the EAA movement. The organization has special categories for Antique and Classic Aircraft, the former consisting of restored planes built prior to 1946, and the latter between 1946 and 1950. Then there's the Warbird Division, consisting of EAA members who have restored military planes built from 1937 until the present.

Dr. Paul MacCready's Gossamer Condor had unique control system for making turns.

For those who love aerobatic flying there's also an IAC (International Aerobatic Club) Division, founded to promote safe grass-roots aerobatic flying. Membership in this division offers educational and instructional materials, and establishes rules for conducting regional competitions.

The big attraction at Oshkosh, of course, is what goes on in the sky. Here, in America's heartland, a dazzling variety of unusual aircraft parade before spectators in the flyby pattern, often against a thrilling background of towering cumulus clouds. Sometimes, a summer shower pelts Wittman Field, but that doesn't stop the show! The crowds simply move in under the wings of parked planes, and keep their eyes glued to the sky until the rainbow comes out.

Safety first is the name of the game, and at the start of each day's program all participating pilots get a thorough briefing on the traffic pattern in use, with the flyby ships divided up into different speed ranges to avoid overrunning. Inevitably, accidents do happen, but Oshkosh's record for safety is outstanding, thanks to the teams of FAA air traffic controllers, ground flagmen, and the squads of EAA personnel charged with directing the incredible number of flying machines on the ground and in the air.

Top man at EAA is President Paul H. Poberezny, a former military pilot and founder of the homebuilt movement, who started the ball rolling back in 1951. Poberezny's lovely wife Audrey amazed everybody at the 1978 Oshkosh Fly-In by putting on a breathtaking wing-riding performance, while son Tom Poberezny, an aerobatic pilot, doubled as convention chairman. The EAA had its beginnings at a Milwaukee suburb, Hales Corners, where today the EAA Air Museum exhibits some 185 historic aircraft, scores of engines, propellers, and other components, plus thousands of photos, models, etc. The EAA Air Museum, incidentally, now carries the name of EAA's President—the Paul H. Poberezny Air Museum.

In January, 1979, an active group of Florida EAA Chapters staged their Fifth Annual Sun N Fun Fly-In, drawing more than 25,000 visitors to Lakeland Municipal Airport for a mid-winter gathering of homebuilt aircraft, while out on the West Coast, in May, more than twice that many people attended the Fifth Annual Sport Aviation Fly-In at Chino, California.

These two events are among a growing number of regional meetings of homebuilt aircraft owners who represent a mushrooming segment of grass roots aviation. The appeal is both economic and recreational—you can build your own aircraft for a fraction of what a factory ship of similar performance would cost, and more, you can build it to suit your personal taste.

Wittman Field, Oshkosh, WI is scene of annual gathering of Experimental Aircraft Association conventioneers, with as many as 8,000 planes attending.

Steve Wittman's fast little W-8 Tailwind can fly at 170 mph.

Many homebuilt aircraft can be flown with far greater economy than store-bought planes, a big factor in a time of growing fuel scarcity. And while some homebuilts may qualify as executive aircraft used in business transportation, the majority are, of course, strictly recreational vehicles of the air, or RVA's.

The Experimental Aircraft Association, along with its sister group, the Aircraft Owners and Pilots Association (AOPA), which represents the bulk of General Aviation's fleet of some 200,000 active aircraft, today work together toward a common goal—maintaining safety and defending freedom of flight in a time when government regulations threaten a takeover of those vast reaches of open sky known as uncontrolled airspace.

Both groups share with the Federal Aviation Administration a concern over safety in flight as we enter the decade of the 1980's. In this decade, an estimated 300,000 General Aviation aircraft will be flown more than 60,000,000 hours a year by more than 1,122,800 airmen. Homebuilt aircraft are still a small segment of this population, with an estimated 9,000 flying in 1980 and another 12,000 under various stages of construction in home workshops.

With more than 300,000 miles of radar-monitored airways providing safe corridors for the bulk of this traffic, homebuilt planes operating on these routes must carry a growing number of expensive and sophisticated avionics devices such as transponders, encoding altimeters, and distance measuring equipment (DME) in addition to navigation and communication radios and emergency locator transmitters (ELTs).

Some homebuilts are able to fly at altitudes of 18,000 to 20,000 feet, to take advantage of the thinner air and substratospheric winds for better speeds on long flights, and to do so must be properly equipped to operate within the positive control airspace where jet transports fly. Recent efforts by the FAA to lower the floor of this positive control airspace to 10,000 feet over much of the contiguous United States were promoted, curiously, by the tragic midair collision between a passenger jet and a private aircraft in an airport traffic area near San Diego, California. The plan was dropped later.

Despite this threatened shrinkage of the free sky there remains a vast airspace where one can fly his homebuilt aircraft with minimal avionics equipment. While sophisticated homebuilts like the Thorp T-18 and Mustang II may range far and wide in the high sky at speeds close to 200 mph, simpler machines, like the growing number of ultralights, fly lower and slower, following a winding river . . . soaring along a mountain slope . . . lazing above a curving coastline. It's a way to see America at its loveliest, with the added pleasure of doing it in a craft built with your own hands.

Chapter 2
You'll Never Get Off The Ground

Samuel Pierpont Langley, a man with no formal college education, firmly believed that flying machines were possible, at a time when aircraft inventors were generally ridiculed as cranks. To Sam Langley, the secret of flight lay in the atmosphere itself, if one could only learn how to harness the power he described in a treatise on "The Internal Work of the Wind".

A noted astronomer and director of Allegheny Observatory in the 1880's, Langley's main interest lay in the fundamental problem of meteorology—the amount of heat the earth receives from the sun. He traveled to the summit of California's Mt. Whitney, 14,495 feet above sea level, measured solar energy levels there, then, satisfied, returned to Allegheny Observatory and began a series of qualitative experiments on lift and drag of airfoils whirled at varying speeds at the end of a rotating arm.

In 1887 Sam Langley became secretary of the prestigious Smithsonian Institution in Washington, D.C., and there conducted more aeronautical experiments with close to 100 rubber-band powered model planes. Satisfied that he had achieved proper stability and balance with his models, he turned to the problem of motive power, in the manner of today's inventive homebuilders, to drive still larger model aircraft. He tried gunpowder, hot water, gas, carbonic acid, compressed air, and electricity, and finally settled on steam.

In the fall of 1891 Sam Langley built his first steam-powered "Aerodrome" and designated it Number Zero. Its hull was patterned after the mackerel, which he felt had the proper streamlines. His

Professor Samuel P. Langley's steam-powered Aerodrome model made amazing flight in 1896 of more than a mile.

engine developed a full horsepower for 41 seconds with 100 pounds steam pressure at 720 rpm, but it didn't get off the ground.

On the West Coast, the *San Francisco Chronicle* devoted a full column to the experiment in March, 1893, in a story headlined:

LIKE A MONSTER FISH

* * *

Uncle Sam Working On An
Airship

* * *

A Smithsonian Scientist's
Invention

* * *

Tiny Engines In A Structure Of
Steel and Aluminum—Wings
Of Silk

* * *

A week later the Chronicle reported that "early last July the first trial was made. No one was present besides the Professor and the constructors. Just as the engines were about reaching their maximum power the craft swayed hopefully, and a cat with a young litter of kittens strolled into the room. Some one proposed to place one of the tiny kittens in the machine as its first passenger. The Professor assented. But the machine would not budge. The kitten was lifted out. Still, the gross weight appeared to defeat the wonderful little engines' power."

Langeley did not give up. He built four more model Aerodromes, with little success, then in 1894 built Number 5, and rebuilt Number 4 as Number 6. He needed more space now to launch his heavier-than-air powered machines, and arranged with the Superintendent of the Coast Survey to use a large houseboat anchored in the Potomac off Chopawamsic Island, near Quantico Station.

Langley invited his old friend, Dr. Alexander Graham Bell, to witness a test flight of Aerodrome Number 5 the following year. Bell had been a firm believer in mechanical flight since boyhood and readily accepted the invitation. Fuel problems aborted the scheduled flight of Aerodrome Number 5 in May, 1895, and further tests were postponed until June.

A.M. Herring a former glider rider who had run flight tests for designer Octave Chanute, was appointed director of field trials, but nothing much happened until the next Spring, when on May 6, 1896, Langley, Bell, and two mechanics took Aerodromes Number 5 and 6 down to the houseboat. At 1:10 p.m. Aerodrome Number 6 was placed on the launch ramp, wings adjusted and motor warmed. It whirred off the houseboat and plunged into the Potomac. A guy wire had snapped and a wing was broken.

Excitedly, Langley and Bell got out Aerodrome Number 5 and gave it a careful preflight inspection. At exactly 3:05 p.m. she was launched. The little craft faltered, dipped three feet, then leveled off and began a steady climb, circling upward in wide spiral. As the men watched open-mouthed, the world's first successful powered heavier-than-air craft purred onward and upward at an angle of some ten degrees, at an estimated speed of 20 to 25 mph.

Later that day, Bell issued an historic statement to the press: "For the first trial," he said, "the apparatus, constructed of steel and driven by a steam engine, was launched from a boat at a height of about 20 feet from the water. Under the impulse of its engines alone, it advanced against the wind and while drifting a little, and slowly ascending with remarkably uniform motion, it described curves of about 100 meters in diameter; till at a height in the air which I and sank into the water, which it reached in a minute and a half from the start. It was not damaged, and was immediately ready for another flight.

"In the second trial it repeated in nearly every respect the action of the first, and with an identical result . . . It seemed to me that no one could have witnessed these experiments without being convinced that the possibility of mechanical flight had been demonstrated."

Had Langley been content to rest on his laurels at this point, he might have been spared future disappointment and yet remained an outstanding pioneer in aviation. He had not only formulated new aerodymanic laws and tables by laboratory experimentation, he had proven them correct with actual field trials.

Flushed with success, he determined to push on to the final goal—construction and flying of a machine capable of carrying a man through the air, guided at the will of the pilot.

Charles M. Manly, an assistant, later said of his chief's decision: "Although in 1896 Mr. Langley had made the firm resolution not to undertake construction of a large man-carrying machine, the longing to take the final great step became irresistible."

Langley was further encouraged by President William McKinley, at whose direction the War Department's Board of Ordinance and Fortification requested him to proceed to develop a practical warplane. The need for such a machine was underscored by outbreak of the Spanish-American War, after the battleship Maine had been blown up in Havana Harbor in February of 1898.

Requesting the sum of $50,000 to develop such a war machine, Langley set to work building an enlarged version of his successful model Aerodrome, with alternations made to permit the pilot to steer it. Searching in vain for a suitable engine, he finally persuaded Manly to accept the challenge. Manly's engine was a work of genius—a five cylinder radial type internal combustion engine, it delivered more than 60 horsepower at 950 rpm, weighing only 3.6 lb/hp complete.

The Langley machine weighed 730 lb and its two tandem wings had a spread of 1040 sq/ft. It was taken to the houseboat, which had been moved to Widewater, Virginia, for launch from a catapault on October 7, 1903, with Manly at the controls. Said the *Washington Post*: "A few yards from the houseboat were the boats of the reporters, who for three months had been stationed at Widewater. The newspapermen waved their hands. Manly looked down and smiled. Than his face hardened as he braced himself for the flight, which might have in store for him fame or death.

"The propeller wheels, a foot from his head, whirred around him a thousand times a minute. A man forward fired two skyrockets. Then came an answering TOOT! TOOT! from the tugs. A mechanic stooped, cut the cable holding the catapault; there was a roaring, grinding noise—and the Langley airship tumbled over the edge of the houseboat and disappeared into the river, sixteen feet below. It simply slid into the water like a handful of mortar."

Investigation by Major M. M. Macomb, Artillery Corps, U.S.A., the official War Department observer, found the cause: "The front guy post caught in its support on the launching car and was not released in time to give free flight, as was intended, but on the contrary, caused the machine to be dragged downward, bending the guy post and making the machine plunge into the water . . . "

Manly tried again on December 8, 1903, but again something went wrong—just as the craft left the houseboat the rear wing and tail assembly crumpled, and the craft slid tail first into the cold Potomac waters. The engine sputtered and died, and with it died Langley's hopes.

The press heaped ridicule on Langley, and even Ambrose Bierce joined the hecklers: "I don't know how much larger Professor Langley's machine is than its flying model," he wrote, "about large enough, I think, to require an atmosphere a little denser than the intelligence of one scientist, but not quite so dense as that of two."

Less than ten days later, the triumphant news of the Wright Brothers' success at Kitty Hawk came as a shock to the aging professor, who died three years later, on February 27, 1906. Said his friend and admirer Octave Chanute: "There is no doubt that the disappointment shortened his life and brought on the attack of paralysis which ended his days."

If Langley died of a broken heart, his full-size aerodrome would eventually fly, 11 years after the Potomac disasters, when Glenn Curtiss secretly modified and retested the machine, added pontoons, and got it off the water at Hammondsport, New York, on June 2, 1914.

The era of powered flight was under way. By 1910 the Wright Exhibition Company, the Curtiss Exhibition Company, and the Moisant International Flyers were putting on airshows across the land. Spurred on by such performances, flying machine inventors came out of hiding and the homebuilt aircraft movement was launched. The first published set of plans appeared in 1907 in *Popular Mechanics*—a primitive Chanute-type biplane hang glider with a span of 20 feet, the spruce spars and ribs covered with muslin. Instructions for flying the glider included a description of how to use body shift to change the center of gravity, a technique used today by modern hang glider riders.

Plans for a monoplane carrying an engine appeared in the same publication in 1910—for only $2 the homebuilder could purchase blueprints for "the world's smallest flying machine" called the Demoiselle, designed by the Brazilian sportsman Alberto Santos Dumont. Demoiselle measured 18 feet from wingtip to wingtip. Made of bamboo sticks covered with muslin, it weighed 242 pounds empty, about the same as a modern ultralight powered hang glider. The engine had two cylinders and produced 30 horsepower. It swung a 72-inch propeller, so heavy that it produced strong gyroscopic forces that made the craft tricky to fly. Unlike the Wright Brothers, who sued everybody in sight to prevent them from building competitive aircraft, Santos Dumont generously announced that he was "placing his invention at the disposal of the world in the interest of the art to which I have devoted my life."

It was a direct slap at the Wright Brothers—Orville and Wilbur were exploiting their Wright Flyer to the hilt, and by the time World War I broke out were deeply involved in litigation trying to prevent others from building and flying aircraft employing the three-torque control system.

Curiously, the basic Wright patent covered the wing-warping method of lateral control, whereas the aileron system, patented in 1911 by Alexander Graham Bell in the name of members of the Aerial Experiment Association, was predated by a similar patent granted May 18, 1897, to Louis Pierre Moulliard of France. The whole thing finally was resolved in 1916 by the United States Congress to get on with the war—a patent truce was ordered, and $640 million was appropriated to build warplanes.

The famous air battles over the Western Front were fought with British, French, Italian, and German planes, not American—by November 11, 1918, when the Armistice was signed, only 740 planes and 77 balloons of American manufacture had reached the war zone. However, a vast pilot pool of Americans existed—11,000 officers and 120,000 enlisted men were in flight training when the war ended.

If the first global conflict stimulated aircraft production and lifted flying machines from the realm of experimental and recreational use, it also precipitated the country into a postwar aviation boom by flooding the market with thousands of war surplus trainers, like the JN4D and Standard, that could be purchased in crates, brand new, for a few dollars. With a ready-made pilot pool of returning veteran airmen, the era of "grass roots" or "gypsy" flying barnstormers was born and would flourish.

This glut did not deter development of more practical sporting planes, however—the ex-combat flyer could purchase a $2500 Ace biplane from Aircraft Engineering Corporation, or the Curtiss Seagull.

However, competition from the government warehouses, chock full of surplus military planes, held the sales of such newcomers to a bare trickle and opened the doors to the homebuilt aircraft movement. Ingenious designers soon were coming up with novel new ideas for personal aircraft. In Los Angeles, George D. White fashioned a lovely little homebuilt, the White Monoplane, which featured a canard-type empennage unit, placed up front instead of in back.

White had hit on something that would provide guidelines for homebuilders for the next half century—build it small, build it cheap, build it in your own backyard, with simple home workshop tools. The White Monoplane was followed quickly in September 1919 by a new homebuilt design called the Penguin, plans for which appeared in *Popular Mechanics.* It was actually a spinoff from a military ground trainer that was not intended to fly.

The barnstorming era added new zest to the American way of life during the Roaring Twenties and brought flying right to the doorstep of rural America, introducing a whole new generation to aviation—swarms of youngsters who ran out to the cow pasture where a rattling biplane had landed, and the gypsy flyer had set up a base of operations, hopping passengers for a penny a pound. I was one of those kids, and I got my first ride in a beautiful machine called a

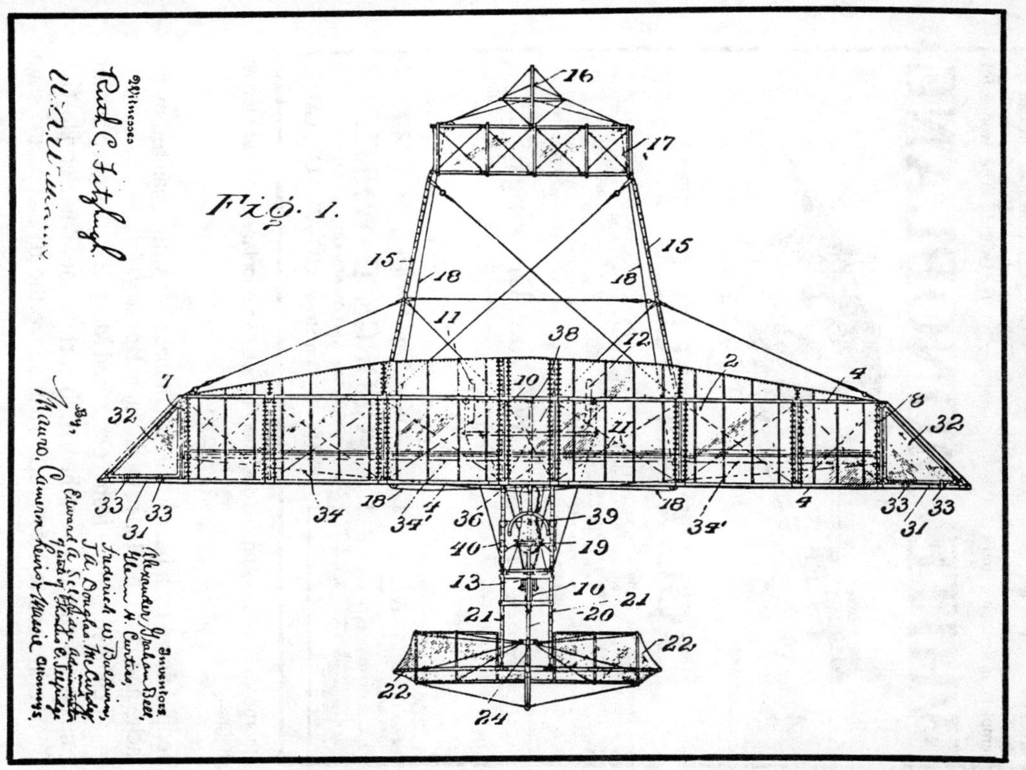

Alexander Graham Bell patented aircraft control system with wingtip ailerons in 1911.

AT LAST!

An Aeroplane for Everybody!

FLIES WITH MOTORCYCLE ENGINE

Think of flying with an ordinary twin cylinder motorcycle engine! This is the only aeroplane that will do it. It is the smallest and most efficient of all aircraft. No longer is flying the sport of acrobats and millionaires. Every man and boy in the world can build one of these remarkable aeroplanes with ordinary tools in a few weeks and learn to fly at home with safety. No shop is needed. If you can use a hammer, saw and a pair of pliers and have a shed, barn, a basement or a back yard you can build one of these remarkable flyers for a few dollars and in spare time if necessary. Costs less than 1/50 the cost of the average aeroplane and can be built for less than a fifth the cost of going to a flying school. It is the smallest, simplest, safest and most successful aeroplane in the world. The wonderful

WHITE MONOPLANE

A Proved Success

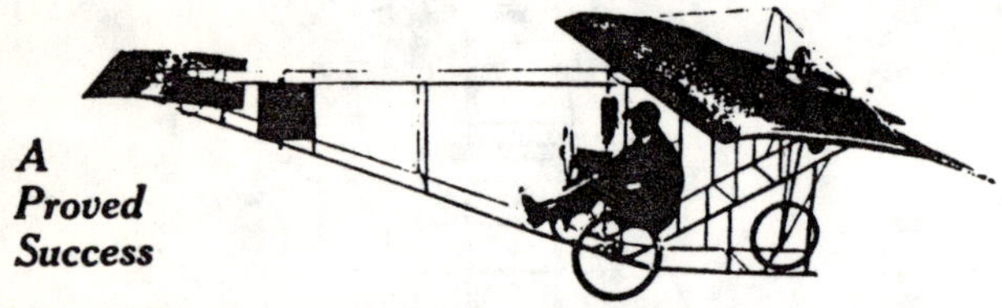

YOU CAN BUILD IT

Remember, this is not a toy or an experiment, but a thoroughly perfected man-carrying aeroplane with 18 foot spread and a speed of 30-60 m. p. h. Lifts 190 pound man with twin cylinder motorcycle engine. Hundreds are already in use in the United States, Canada, Mexico, South America, England, Australia and the Philippine Islands. Hundreds of amateurs are building and flying these aeroplanes, men and boys who knew absolutely nothing about aeroplanes and flying before. It is the simplest and safest flyer in the world. Any make of motorcycle engine can be used.

WHAT OTHERS SAY:

POPULAR MECHANICS says:
"Undoubtedly the smallest successful monoplane in the world."

AERIAL AGE says:
"One of the most interesting machines yet to be developed."

SCIENTIFIC AMERICAN says:
"A unique type of monoplane."

L. A. EVENING HERALD says:
"Has solved the problem of producing small machines at a cheap price for universal service."

Working Drawings $2

Send $2 at once for a complete set of working drawings of this wonderful monoplane showing all details and dimensions in a simple manner so you can easily understand everything. Here is your chance to get into the greatest of all industries. Thousands of experienced flyers and builders are needed.

Don't Miss This Chance! Send $2 Now!

GEORGE D. WHITE
DESIGNER AND OWNER OF SOLE RIGHTS
3832 South Main Street, LOS ANGELES, CALIF.

George White of Los Angeles designed homebuilt monoplane, sold plans for two dollars.

Jenny, flown by a handsome ex-combat pilot with a painted moustache and shiny boots, named Major Hensley.

I climbed into the front cockpit and Major Hensley gave her the gun. We bounced over the rough turf of the Wheelerville Golf Course in upstate New York, rose above the treetops, and soon were circling our home at nearby Canada Lake, where I'd spent many a summer afternoon lying on my back, looking up at the fleecy cumulus clouds, knowing that someday I would be up there, flying with the birds. My life's course was set—I would build a primary glider in high school, join the Army Air Corps to fly military gliders, and later become a flight instructor with the Royal Air Force, flying Stearman PT-17s, in the second World War. Peace would come again to the land, and once more you could buy a war surplus training plane, brand new, for only $600, with a full tank of gas. A rerun of the 1920s.

The second big stimulus to backyard aviation between the world wars was, of course, the great flight of Charles A. Lindbergh from New York to Paris in 1927. We all sat by the radio (no television then, remember?) listening to the news of Lucky Lindy's progress, praying that he'd make it. He did, and a new hero was born, one who would leave an unforgettable impression on young minds.

All that is history now, along with the scores of little homebuilts that took to the sky between the big wars. But today, in a world beset by more confusion and pressures, history seems to be repeating itself—again the sky is filling with restorations and replicas of the homebuilts of the 1920s and 1930s. Whenever you attend an EAA flyin, you'll likely see a few of these graceful dragonflies darting about overhead—Heath Parasols, Corben Baby Aces, Pietenpol Air Campers. In a later chapter we'll go more deeply into the fun of restoring and replicating yesterday's planes, but for now, turn the page and read all about a modern-day Baby Ace, and what it's like to fly one.

Chapter 3
Flying the Corben Baby Ace

To the Corben Baby Ace goes honors for launching the world's foremost homebuilt aircraft fraternity—the Experimental Aircraft Association, which today numbers some 120,000 members. This happened after EAA's President Paul H. Poberezny, a former World War II combat pilot, acquired the rights to build the Baby Ace, commissioned the late Stan Dzik to draw up a set of plans, and got it all together as an exceptionally clean little sport plane, for one.

Popular Mechanics magazine, again intrigued by the lure of tiny homebuilt planes, ran an article on the EAA version of the Baby Ace, along with a set of plans. The response was unexpected—hundreds of Baby Aces soon were flying, behind a variety of engines. The EAA, then a handful of midwestern homebuilt freaks, quickly won national attention.

Back in a corner of the Paul H. Poberezny EAA Air Museum near Milwaukee sits the original Baby Ace that started it all. The ship had been sold, then bought back for the Museum, because of its historical importance. Its engine is a Continental A65, the landing gear Cub-type, but basically it's the same ship, restored, with its original color scheme.

It's your guess how many Baby Aces are around today—some 70 are believed actively flying in the United States and Canada, and hundreds of sets of plans have been sold by Ace Aircraft Mfg. Co., 106 Arthur Road, Asheville NC 28806.

One day I flew up to Mojave Airport, on the desert north of Los Angeles, and ran into Bob Upton, a Rocketdyne patent agent, who, back in 1972, found a bare-bones fuselage of a Baby Ace and bought

Corben Baby Ace started the Experimental Aircraft Association movement after Popular Mechanics ran set of plans.

it. He trucked it down to Oceanside Airport near San Diego where Paul White, a veteran aircraft builder, completed its restoration.

Upton's Baby Ace shared a hangar at Mojave with Joe Mason's Baby Lakes, a tiny sport biplane, until Bob had flown off the first 75 hours. Then he could legally move it down to Van Nuys Airport, closer to home. It was there Bob checked me out in his Baby Ace, simply by standing beside me in the cockpit and going over the speeds and numbers and then saying, "Okay, Don, go fly her!" It's that simple to fly —no big mystery.

First, though, let's do a walk-around inspection. Baby Ace has the same pretty lines of two other popular parasol planes today—Poberezny's Pober Pixie and, of course, Ed Heath's Parasol. Ace Corben was a country boy of the same mold as Heath, Bernard H. Pietenpol, and other early aircrafters who got grass roots flying into the air during the Great Depression years.

The original Baby Ace was a true ultralight, with an empty weight of only 400 pounds and a 300-pound useful load 63% of the machine's entire weight. The Clark Y wing spanned 25′ 4″ and she stood 17′ 5″ in length from spinner to tail.

In 1931 the Corben Sport Plane & Supply Co. of Peru, Indiana, began manufacturing the Baby Ace either as a kit plane or complete as a flyaway unit. Kits included the fuselage, landing gear, controls, and tail unit, all prewelded at the factory.

There were two models, the parasol monoplane and a high-wing version with a passenger cabin and wider span of 27′ 8″; its empty weight 475 pounds. Spruce spars and ribs were used, and the parallel wing struts were of steel tubing, resembling those of the Model B Funk which appeared later.

Looking something like a giant model plane, the Baby Ace uses balsa wood extensively for fairing and streamlining. Wings, tail, controls and landing gear were interchangeable with either the open or closed cockpit version. The parallel wing struts eliminated the problem of interior wing rigging and were easily removable so the wings could fold back for road-towing. A detachable motor mount permitted the builder to switch engines with no major reconstruction involved.

Corben offered a wide range of engines—the Szekeley SR-3 of 45 hp.; the Continental A40 of 35 hp.; the Salmson AD9 of 40 hp.; and the Heath B4 of 30 hp. The landing gear was of the split-axle type, like the J-3 Cub, with bungee shock absorbers. Two other versions appeared—the Model E Baby Ace, a side-by-side two-seater, and the Super Ace that carried a Model B Ford engine.

In 1955 Poberezny redesigned the original Baby Ace as the

Model D, and most flying today carry the Continental A65 engine. Plans are available for the single-place Baby Ace and for the two-place Ace Junior from Thurman Baird, 106 Arthur Road, Asheville NC 28806. Progressive kits also are available.

Bob Upton's Baby Ace is a bit heavier than the original model, weighing 612 pounds empty. Wing span is 26′, length 20′, height 7′, wing area 112 sq/ft. Upton installed a 12-gallon fuel tank forward of the cockpit area, using a Cub-type wire float gauge sticking up over the cowl. In economy cruise mode this provides a range in excess of 350 miles.

A roomy right side door provides easy entry to the cockpit—you just swing a leg in, grab a cabane strut, settle down, buckle seat belt and shoulder harness, and slide your feet onto the rudder pedals. Aeronca-type heel brakes are there if you need them.

I felt comfortable sitting in the Baby Ace, getting used to the view and scanning the panel, neatly laid out in a handsome varnished wood mount. Airspeed indicator at left, sensitive altimeter to the right, then a magnetic compass, skid and slip indicator, and tachometer. Beneath were an 8-day clock, oil temperature and pressure gauges, and the ignition switch at far right. An Escort 110 nav/com unit was mounted low between the rudder pedals, powered with nickel cadmium batteries. A vernier throttle was handy at left, under the carburetor heat control.

Little slipstream buffeting appeared as I taxied for takeoff, and a flat-sided windscreen had a sort of World War II military appearance. On takeoff, the tail came up easily with the control column in neutral position, and liftoff came in roughly 400 feet at 60 mph IAS for a Vx climb at 65 mph of 750 fpm to pattern altitude.

There was a definite feeling of *deja vu* as I left the pattern and climbed out to 3000 feet in a practice area to run some stalls and get the feel of this born-again beauty. With throttle set at 2000 rpm and the needle on 65 mph IAS, the best climb rate speed, I raised the nose and was surprised to find she didn't want to stall at all. I finally got a shudder and then a break, with recovery straight ahead and smooth, no roll off tendency. I'd gained 300′ altitude.

Power off, I trimmed for landing attitude and stalled her straight ahead, left, and right. Same thing—an honest stall warning with slight buffeting over the Clark Y curve. Full stalls did reveal a minor rigging problem, as the Baby Ace insisted on falling off on the right wing, not wildly, but noticable.

The ailerons were surprisingly heavy, like those of a VP-1, requiring heavier than normal pressure to enter or leave a turn, but directional stability was excellent—boot in left or right rudder, take your foot off, and back she comes to straight and level in a couple of oscillations.

Happy with knowing her whims, I spent the next half hour admiring the view—excellent in all directions but straight up, where the parasol wing interferes, especially in turns. I suggested to Upton later that he cut a hole in the wing and install some transparent plexiglass (he did).

Careful pilots, however, do not enter turns in high-wingers without first clearing the area with a healthy look-around. By raising the wing before entering the turn, you can be assured the sky is clear of traffic, just like flying a Cessna 170 or a Luscombe.

There is a certain thrill in open cockpit flying you don't find inside the enclosed cockpit of a modern factory job—the smell of fresh air, the improved visibility, the sense of being a part of the sky, not driving through it in a capsule. Baby Ace has a charisma all her own. She's an ultralight, but not too light; responsive, but not tricky. Properly trimmed she'll do 100 mph IAS at 2350 rpm, or cruise along economically at 80 mph IAS on 2000 rpm to save gas, if you're in no hurry.

Returning to land, I found an approach speed of 65 mph IAS with a touch of power provided good forward visibility, although 60 indicated worked about as well. She stalls around 38, hence you can make a safe short-field approach down in the 50's. Flareout blocks the forward view, but shifting your eyes a bit left or right provides plenty of visibility to make a three-pointer. On rollout, the rudders are a bit sensitive, so don't go to sleep until you're tied down!

Chapter 4
Pietenpol Air Camper

1930 was a vintage year, despite the Great Depression, Prohibition, Hoovertowns, and the mysterious disappearance of New York State Surpreme Court Justice Joseph F. Crater. Bathtub gin was a buck a quart, the Lindbergh boom was over, and you could go to the talkies for two bits and catch the latest Fox Movietone News, after Clark Gable and Myrna Loy.

The Dodgers were still the Brooklyn Bums, Herbert Hoover was still in the White House, and if you owned a raccoon coat you could shake out the mothballs, hop into your Model A and fly into town for a date with your favorite flapper, to forget the big stock market crash that wiped out Uncle Percy.

Fly into town in a *Model A?*

That's right—back in a time when there was no TV and folks listened on the radio to the Happiness Boys, Amos and Andy, and Russ Columbo, the Model A Ford was every schoolboy's dream car. Its four-cylinder, water-cooled engine could grind out forty horses on gas that cost only 17 cents a gallon. The engine was such a success, in fact, that *Modern Mechanix* Magazine featured it as the powerplant for a kooky little homebuilt airplane called the Air Camper, a genuine flying machine that could go faster than a Model A car, fly 200 miles from the farm into town and land in the local ballpark.

The do-it-yourself craze was just getting rolling as a Depression-born phenomenon, and the Air Camper replaced primary gliders as homebuilt aviation's favorite backyard project. Designed by Bernard H. Pietenpol, of Spring Valley, Minnesota, it quickly moved into first place among amateur-built aircraft, offering safety, simplicity of construction, and low cost.

Pretty soon, the skies were filling with staggering Air Campers, some of which performed better than others, depending on how well the builder followed the plans. If he was a completely inept do-it-yourselfer, he could still write to the Pietenpol factory and buy one fly-away for around $750. No flying field was complete without an Air Camper chugging around the pattern, and many others were flown from local farms by young eager mechanics who simply tied them to a tree at night, or pushed them into the barn when the winds blew.

B. H. Pietenpol's name was as well known to youngsters back in the Thirties as that of Bill Piper, who got his start in aviation when he bought out C. G. Taylor's E-2 Cub and made the name a household word. But Pietenpol had no interest in cranking out Air Campers on a production line basis. A typical Minnesota farm boy, he just wanted to get off the ground and have some fun in the sky. He'd learned to fly back in 1922 in a war surplus JN4D Jenny, but gave up the idea of owning one because of the maze of wires and struts that meant multiple maintenance headaches.

One day I asked Pietenpol how he came to design and build the Air Camper. He replied cryptically: "I wish I knew myself. I often wonder about that!" Then he elaborated a little: "I just liked to build things, and I wanted to fly, and so I decided to build. There weren't any ships to buy in the Twenties, except the Jennys."

It was back in 1923 that Pietenpol assembled his first homebuilt, a Model-T powered biplane. Next he built another bipe with a Gnome rotary engine installed. A Lincoln Sport followed, this one with a Lawrence engine. At that time, Pietenpol was running a Ford agency in Cherry Grove, Minnesota, and he made up his mind that the simple Ford engine was about as reliable as any aircraft powerplant in the same price range—Ford cars sold for as low as $300 brand new, remember?

So it came about that Pietenpol revised his whole philosophy of homebuilding in the interest of simplification—he'd use only a single wing instead of two wings, and get rid of the bird's nest of bracing wires in the fuselage in favor of the simple Warren truss wood construction. By 1929 he had three monoplanes flying with Ford engines up front, and was happily poking along on cross-country hops to visit relatives as far away as Winona and Rochester.

Came the 30's. One day Pietenpol picked up a copy of *Modern Mechanix* magazine and was astounded to read that automobile

Bernard H. Pietenpol was a leading homebuilt designer of the 1920s and 1930s with his popular Air Camper.

engines would never become a cheap substitute for aircraft powerplants. He got out pen and paper and wrote an indignant letter to the editors in St. Paul, offering to fly his two automotive-powered planes to their office and put them straight. The editors met him at the airport and stared in disbelief. They were so delighted they pleaded with him to draw up a set of plans, so their readers could share the fun.

So it was that the 1932 edition of *Modern Mechanix Flying and Glider Manual* ran a complete story on how you could build your own Depression flying machine, which the editors labeled the Air Camper. It carried the Model A Ford engine, a four-cylinder hunk of iron that put out 40 horsepower, with a big water-cooler radiator up front.

To help out the farm boys trying to get along on Depression dollars, the 1933 edition of the same publication came out with plans for a Pietenpol Sky Scout, powered with the earlier Model T engine that had gone out of production in May, 1927, after more than 14,000,000 had been built and sold, a record that stood for 44 years until, in 1971, the Volkswagen Beetle became the new champion and provided a new automotive engine for a future generation of homebuilders.

Pietenpol eventually drifted into commercial aviation as a full time occupation, became a World War II flight instructor, and finally

Tom Baker working on his Pietenpol Air Camper.

settled down again in Cherry Grove, fixing new-fangled TV sets for a living and building up more Air Campers for the hell of it.

In 1966 Pietenpol found a used Corvair engine and built another Air Camper around it, 9 inches longer and with 4 inches more wingspan. She cruised at 80 mph, could hit 100, and fly 5-1/2 hours on 19 gallons of go juice.

His last Air Camper was flying in 1972 with the center section braced by inverted-V and Jerry struts on the wing, and he fancied her up with a fiberglass cowling, the exhaust stacks protruding through the bottom to keep her quiet. In a fitting tribute to a great guy, the EAA in 1975 awarded Pietenpol first prize for the best automotive-powered homebuilt—his Corvair-engined Air Camper, N7533U.

Many Air Campers flying today have abandoned the old Ford engines for Continentals and Lycomings, but purists believe it isn't a

Some Pietenpol Air Campers Flying today use modern aircraft engines instead of Model A or Model T Fords.

Framework of Tom & Lee Baker's Pietenpol Air Camper.

Pietenpol if there isn't a Ford under the hood. Doug and Maryann Knode, of Santa Clara, California, wouldn't think of having anything so terrible as an air-cooled aircraft powerplant in their Air Camper, which consistently wins ribbons at West Coast Fly-Ins. They named their ship N3133, which happens to be the same as their house street number, and it's one of eight flying today behind 60 horsepower Model B Ford engines.

Their Pietenpol isn't the world's fastest, yet it cruises along at a good 75 mph at 1850 rpm and can hit 85. She'll settle onto the runway in a three-point stall landing at a gentle 38 mph. They installed a 12-gallon fuel tank in the wing's center section so that the gas gravity-feeds into a Stromberg carburetor, and behind the engine is hung a big radiator that holds two gallons of water. Doug fitted a wrap-around heater manifold that takes ram air over the front exhaust stack and feeds it through flexible tubing into the air intake, making a simple but effective carburetor heater.

Homebuilt Pietenpol Air Campers differ in detailing according to the builder's whim, but some items are common to all of them, like the split axle originally used by the designer. In his 1972 model, Bernie used special shock struts made from an Aeronca Defender landing gear spring cut in two. He claimed it picked up 2 mph in cruise speed over the shock cord strut version, and while it gave a good ride he felt it should have a bit longer spring length.

The wing curve was Pietenpol's very own. At first he tried a number of airfoils then in use—the Clark Y and various NACA sections—but all required a long takeoff roll with the Ford Model A engine. One night he and a buddy sat down at the kitchen table over a cup of coffee and doodled out a new curve that pleased their eyeballs. It flew fine.

Another goodie Pietenpol added was the angle of the engine installation; it provided some downthrust, which he claimed helped the propeller bite the air more squarely in climb attitude.

Of all the Pietenpols flying today, one with perhaps the most romantic story was built by a young pair of newlyweds, Tom and Lee Baker. Tom had enlisted in the Army in 1972, and while stationed at Manhattan, Kansas, the couple sent for a set of plans and set to work in their upstairs apartment, starting with the fuselage, which they assembled in their bedroom.

Later on, the project took over most of the rest of their living space in kitchen, bathroom, and living room. When Tom was trans-

Tom Baker pulls through prop on his Air Camper.

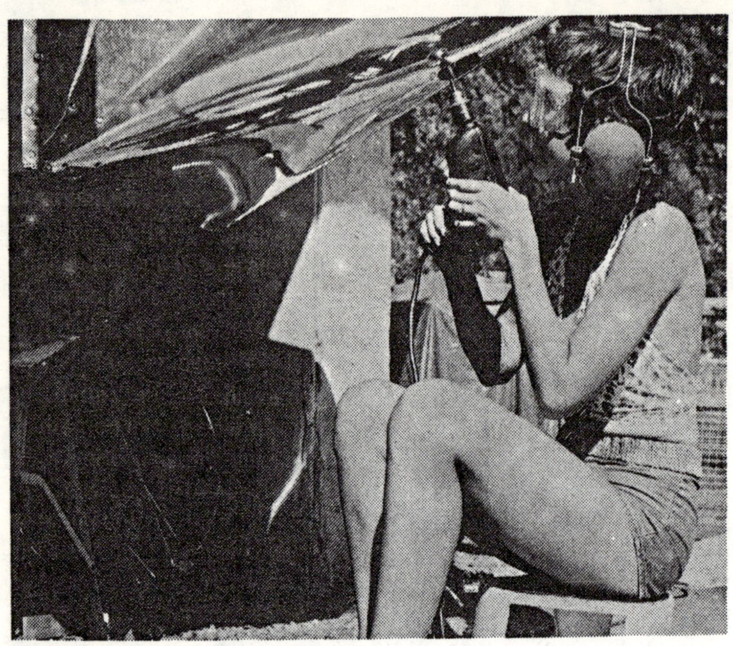

Like a good wife, Lee Baker gives a hand at building Air Camper.

Tom Baker visits with B. H. Pietenpol, designer of Air Camper.

ferred to Albuquerque, they borrowed a van and hauled the pieces along. Finally, the fuselage was lowered from their upstairs bedroom by rope, the wings, tail, and landing gear added, and the engine and prop installed up front.

On September 15, 1975, their Air Camper was completed, at a cost of some $3000, and three weeks later Tom's dad, Donald A. Baker, a licensed pilot, made the first test hop. The following month a tragedy occurred. A stranger approached Tom, claimed he was a flight instructor, and offered to take the Air Camper up for a spin. He did—and spun in, totaling the bird and causing serious injuries to the unfortunate pilot.

Tom felt bad, of course, and changed engines from the Continental A65 to a Model A Ford, to get more power for flight operations from the mile-high airport at Albuquerque. At this writing, the reconstruction project was well under way. You can't keep a dedicated homebuilder down!

Chapter 5
Flying the Heath Super Parasol

We now come to Edward Bayard Heath, one of three early designers who got the homebuilt aircraft movement moving, the others being Ace Corben and Bernard H. Pietenpol. After developing and flying a number of hot little racers like the Baby Bullet and the Tomboy, Heath was approached by a Minnesotan named Weston Farmer with the idea of developing an ultralight monoplane using a Henderson motorcycle engine and the lower wings from a Thomas Morse Scout.

A considerable number of the resulting ultralight, called the Heath Parasol, was built, before the evolutionary Super Parasol evolved, with a Clark Y airfoil replacing the World War I thin wing of the Scout. By 1929 the Super Parasol was all the rage, and one evening that year a midwesterner named Irvin C. Rofshus settled down to read a magazine before the fireplace, and came across a picture of a helmeted pilot flying a small parasol plane into the setting sun.

"Listen to this, Ma!" he yelled, then he began reading the magic words that would change his life:

HOME ON TIME!

With the sun setting in the west your HEATH SUPER PARASOL driven by that dependable HEATH-HENDERSON aircraft motor brings you home on time.

The HEATH SUPER PARASOL is the most economical transportation ever devised by man, with a fuel cost of less than 1/2¢ per

Ed Heath in his seaplane Parasol.

mile. Build this plane yourself for $199, or purchase it flyaway our Airport at $975.00. Detailed shop blueprints at $5.00 per set. 10¢ in stamps or coin brings you our large illustrated booklet. HEATH AIRPLANE COMPANY, 1721-29 Sedgewick St., Chicago, Illinois.

Rofshus went to his writing desk, dug out five 2-cent stamps stuck them in an envelope, and sent off for the information kit, and before long he was on his way to become one of the first home-builders to put together a Super Parasol.

The Rofshus Super Parasol is still flying today, after half a century of service in the same family. She was restored a dozen years ago, and its pilot-owner now is Irvin's son Robert, a plastic mold machinist of Albert Lee, Minnesota, who already has logged some 200 hours on the antique homebuilt.

The Super Parasol used to advantage the modular construction concept that had launched the earlier Parasol as America's first kit homebuilt and later would make Heath Kits a household phrase in the world of electronics. You paid $12 for the first kit, or, if you were rich, plunged and paid $199 for the whole assortment.

"For those who cannot afford to invest the whole $199," Heath advertised, "we have arranged shipment of the bill of materials for our Heath Parasol in eleven convenient progressive groups. By purchasing these in groups you may build your Heath Parasol with limited capital and without delay."

He added: "If there's a small river or lake near by, remember that you may build the Sea Plane Parasol on the same installment basis."

The original Heath Parasol grew bigger with stronger wings over the years as its horsepower went up, but basically it remained one of the most popular homebuilts ever designed. If you weren't handy with tools, you could still order a Parasol complete, flyaway from the factory for only $895, with a 23-horsepower, four-cylinder Heath Henderson De Luxe motorcycle engine that swung a 4-1/2-foot propeller directly driven on the crankshaft by means of a special conversion Heath had invented.

The two wing panels were pinned together at the center, and so could be quickly disassembled for road-towing and storage in the barn. And if you didn't know how to weld, the steel tubing fuselage could be bolted together with a wrench and pair of pliers.

The 1927 Parasol had a wingspan of 23 feet and weighed 290 pounds empty. It could carry 225 pounds and was capable of a top speed of 70 mph. Landing speed was 32 mph. By 1929, Heath had added two more feet to the wingspan, thus increasing the wing area from 94 to 110 square feet. The empty weight dropped to 260

Bob Rofshus with his Super Parasol.

pounds, and the useful load was increased to 300 pounds. Several engines were available, from the 4-cylinder converted Henderson to the 3-cylinder Anzani, the 2-cylinder Wright Morehouse, and the 2-cylinder Bristol Cherub.

Before his tragic death in a plane crash in 1931, Heath had modernized the Parasol with a left side door for easier entry. V-struts joined at the bottom of the fuselage, with the aileron cables strung through the front struts. Heath added a skylight to the center section for better visibility in turns.

Other modifications included skis and floats, the changeover made simply by using the same fittings that carried the conventional landing gear; and a low-wing model also was introduced. The Parasol design remained the favorite over the years, and the most popular was Model LNB-4, the one granted Department of Commerce approval for home assembly.

LNB-4 came with optional starter, fuselage door, wheel brakes, and a choice of three propellers—linen-tipped spruce, hard wood, or metal. It weighed 450 pounds empty and with the Heath B-4 engine could hit 73 mph. Of course, nobody was really in a hurry in those days. Gas was ten cents a gallon and a dollar's worth would carry you 330 miles. Transportation at one-third cent a mile was not at all bad.

The Heath Super Parasol, introduced at the Chicago Aircraft Show in 1929, was the ultimate ultralight of its time, weighing only 260 pounds empty. Powered with the converted Henderson motorcycle engine of 27 horsepower, she'd hit 70 mph, land at 28 mph, and had a range of 200 miles.

Engine and propeller weighed 117 pounds, about half the aircraft's total empty weight. Wings were of conventional design with spruce spars and spruce ribs and webs. The ribs weighed only 2-3/4 ounces each, but carried a rather heavy load.

Two inverted-V struts in front and back of the single cockpit supported the wing roots. Two steel tubes on either side of the fuselage ran from the bottom longerons to about 60% out along the wing panels. A 5-gallon fuel tank in the center of the wings worked by gravity feed. At 40 mpg this provided a range of 200 miles without reserve. The Super Parsol's wing panels weighed only 18 pounds each, and the wingtips were made from tubing bent into a semielliptical shape. Ailerons were hinged from the top of the rear spar, increasing the control leverage and at the same time making the hinges handy for inspection or servicing. The airfoil was the Clark Y.

Bob Rofshus, being a skilled mechanic, tore down the Corvair engine in his father's Super Parasol three times over the years,

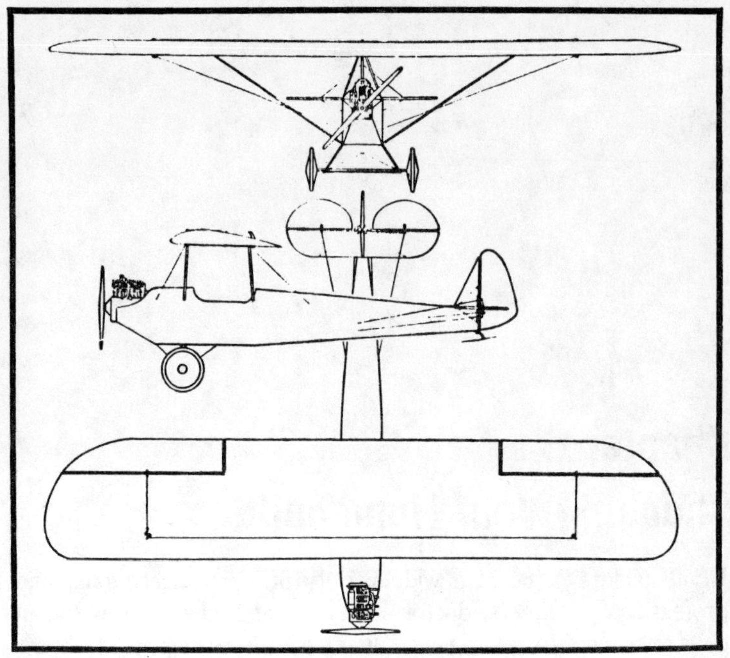

Three-view of Heath Super Parasol.

modified the air baffles, and extended the exhaust manifold stacks to improve cooling. The first time he rebuilt the Corvair with 1/8-inch overbore racing pistons, but after 24 hours she began to rattle and bang, particularly at a high rpm of around 2700, with a 43-inch pitch Al Schauss wooden prop.

On one 80-mile cross-country trip Rofshus pushed the Corvair up to 2850 rpm and hit 100 mph, when suddenly the whole engine began to flutter. He throttled back quickly and things quieted down, and when he finally got home he discovered a big hole in one piston and a hunk of piston ring clanking around inside the exhaust port.

Discussing the problem with an old friend, Bernard Pietenpol, who lived over in Spring Valley some 60 miles east of Albert Lea, they came to the conclusion that burning 100-octane gas had burned the hole in the piston. He switched back to regular automotive fuel, which the Corvair was designed for.

One time Rofshus flew over to the EAA Fly-In at Rockford, Illinois, and slid into the traffic pattern behind Ray Hegy's famous homebuilt, Chuparosa. He hung right on Ray's tail, and Hegy kept peering back over his shoulder, wondering how a Super Parasol was staying with him at 120 mph. It was the Corvair that did it.

Chapter 6
Choosing Your Homebuilt

Homebuilders come in a wide variety of ages, inclinations, and professions—students, doctors, lawyers, and maybe a few Indian chiefs. All have something in common, a born-free spirit. The spirit of adventure, the desire to be creative, the urge to seek fulfillment in freedom of the sky, to be one with the birds.

The majority are practical people, working on a limited budget, both financial and timewise, anxious to turn a hobby into a useful occupation, to beat the high cost of buying a brand new factory plane, to end up with a ship embodying pride of ownership.

Aside from such personal motivations, homebuilders by and large may be highly sociable fellows and gals, sharing their building experience with others who have the same inclination, or they may be "loners" who enjoy holing up in the basement or garage, turning on the TV and watching a ball game while they give substance to their dreams of a perfect flying machine.

There are those who would rather build than fly, who simply enjoy the challenge of turning a blueprint into reality, and when that is accomplished, sell the ship to a buddy and start on something else. Others, of course, are born flyers, whose ultimate dream is to sit in the cockpit of a plane they designed and built themselves, and fly off to faraway places.

There are a large number of ex-military pilots who learned to fly during the war years, and today are building and flying scaled-down versions of the fighters they flew in combat. Many other builders who saw service remember the thrill of their first trainers—

Rutan VariEze is excellent new homebuilt design that utilizes modern composite materials construction.

Stearman PT-17s, Fairchild PT-19s, or whatever—and want to get back in the sky in smaller aerobatic biplanes or monoplanes to run through the old drill—loops, rolls, spins, or whatever—to sharpen their skills and renew the precision techniques of their student pilot days.

Among the younger generation are builders working on limited budgets who want to get started flying in such craft as hang gliders, with or without engines, or other ultralight aircraft that will cost them maybe $1,000 or less. Then there are the nostalgia buffs, who want to recreate the good old prewar days and go skimming across the countryside low and slow in pretty parasols, like those described in previous chapters.

Whatever the inclination, there are two ways to go—either start from scratch and design and build an entirely original aircraft, or better still, select a tried and true design from a professional aeronautical engineer, and follow directions. In the latter category, you can usually get started by sending off for an "Info Kit" for $5 or $10, make up your mind if that's what you want, then send for a set of plans or kits, which often are available in increments, so you don't have to lay out the full amount all at once.

In the back of this book you'll find a listing of the more popular homebuilt aircraft now flying, with specifications, performance figures, addresses, etc., but only you can decide, subjectively, which one is for you. Your first consideration as a serious amateur builder should, of course, be safety—will the design meet your requirements for the kind of flying you intend to do?

It's been said that there's an aircraft for every kind of pilot. Some relish wringing out a ship in precision aerobatic maneuvers, pulling 5 or 6 G's (5 or 6 times the force of gravity) in violent snap maneuvers, lomcevaks, or whatever. Others enjoy riding in comfort, inside a warm cabin, particularly during a cold winter when even the birds are grounded.

There are sleek, low-drag, high-performance cross-country planes that can get you from here to there in a hurry, planes that can carry two couples, and single-seaters for the pilot who likes to get away from crowds and find solitude in the sky.

There are monoplanes, biplanes, and even some triplanes; high-wingers, mid-wingers, low-wingers; even aircraft with no wings at all—the rotorcraft types that use rotating blades overhead to produce lift. In considering monoplanes, a high-winger offers the cleanest view of the landscape below you on a cross-country flight, although visibility suffers in a turn. If you enjoy looking up at the sky, or prefer better visibility flying in airport traffic patterns, you may choose a low-winger. A mid-wing aircraft offers better trim at all power settings, with thrust and drag forces balanced along a centerline.

More important, of course, is the safety factor, for even if your workmanship is impeccable, you want to know that the design is airworthy, in the sense that it has been properly engineered. Has the designer run a stress analysis on the airframe? Has he chosen the right engine and propeller? Does the CG fall in the right place?

Burt Rutan, designer of the lovely VariEze and other sophisticated aircraft built from composite materials, asked his brother, Dick Rutan, a retired Air Force pilot, to conduct a series of flight test maneuvers on VariEze to explore a phenomenon other VariEze builder/pilots had reported encountering—a highly divergent wing rock, or an abrupt rolloff into a nose-down rolling dive.

Dick knew from past experience that VariEze was virtually spin-proof, but that it would execute a "rolling departure" from level flight—an "auger" in military parlance. However, even with full aft stick and full rudder recovery was inherent and rapid. Nevertheless, you could quickly lose 2000 feet of sky in such a maneuver, so Burt enlisted NASA help to run a series of wind-tunnel tests on a VariEze model fitted with a partial-span drooped cuff. The cuff, trimmed to 38 inches in length, eliminated the departure phenomenon and wing rock damped out quickly after excitation.

Dick Wagner's popular Wag-A-Bond is a replica of the Piper PA-15/17 Vagabond, makes a fine flying camper.

Dick Wagner not only designed the Wag-A-Bond, he added a special tent called the Wing-A-Bago for flying campers.

Hundreds of VariEzes are flying today, and the design is excellent. It makes a fine cross-country ship with a high cruise speed, is comfortable and safe, properly flown, and a new model introduced in late 1979, the Long-EZ, added more wing area by virtue of a wider center section, a "rhino rudder" on the nose, and a roomier cockpit. However, the VariEze is not an off-airport sport plane, one you would want to operate from a short dirt strip.

In the latter category, one might turn to a number of fine biplanes like the EAA Acro Sport, the Wichawk, or the Steen Skybolt, but the homebuilder who loves the outdoors and enjoys getting away from the crowded airports for a weekend of fishing in a distant mountain stream could do worse than consider something called the Wagabond.

Dick Wagner, an airline captain who runs an aircraft supply house in Lyons, Wisconsin, likes nothing better than to fly off to

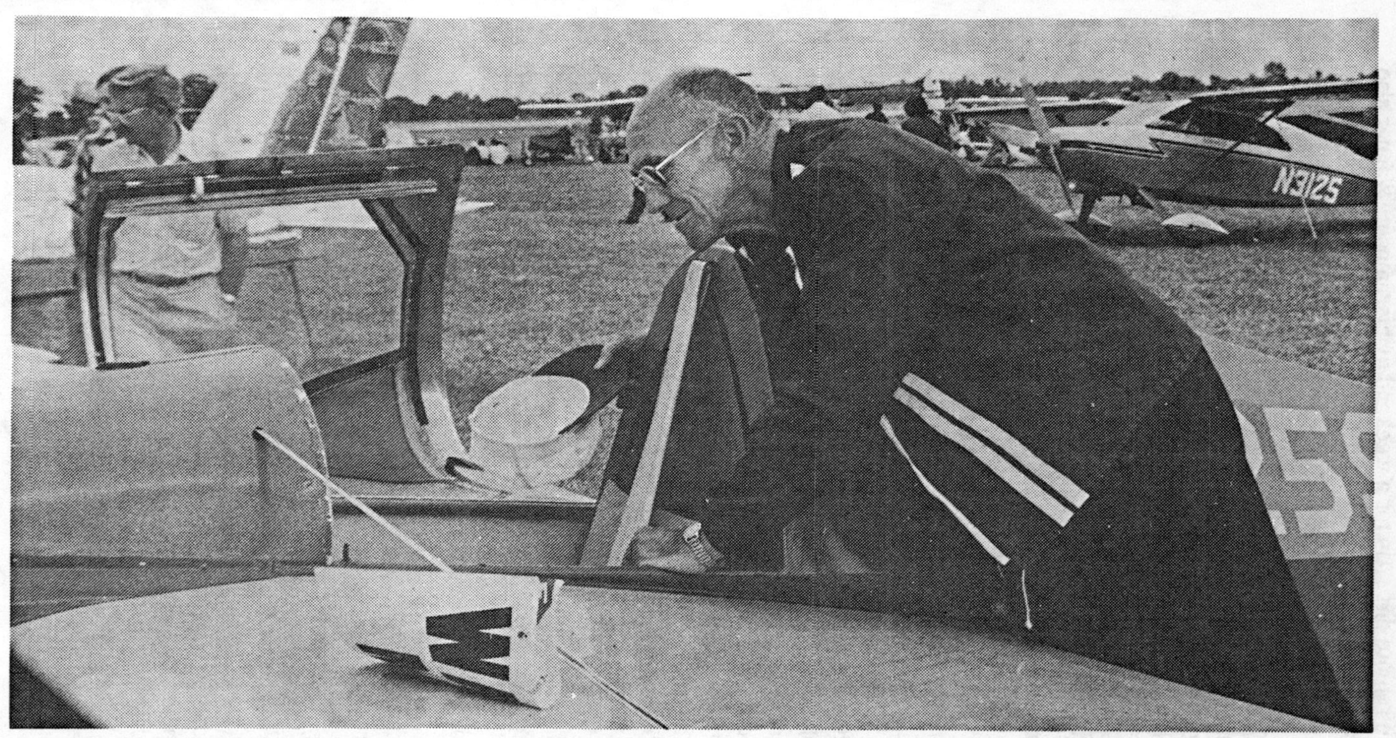

Steve Wittman designed a Formula Vee racer called the Witt's V, uses a VW engine.

some remote dirt strip for a weekend flying campout. Wagner also loves antique aircraft from the 1920s and 1930s, ships designed for use at a time when most airports were in fact dirt or grass strips.

As a consequence, he adopted the design of the postwar Piper PA-15/17 Vagabond as the basis for a new homebuilt he called, naturally, the Wagabond. You build it yourself from kits supplied by Wagner's firm, Wag-Aero, and use a Lycoming 0-235 of from 108 to 115 horsepower, compared to 65 hp in the original model Vagabond.

This gives you a nice cruising speed of around 115 mph, plus good short-field performance. Further, Wagabond's seat folds back to allow pilot and a friend to stretch out inside the fuselage, unrolling a pair of sleeping bags on a soft rubber mat for a good night's sleep. Wagner also developed a line of tents you can slip over the wing of your plane, whether it's a high-winger or low-winger, for more room. He calls them Wing-A-Bagos, a play on the name of a popular road mobile home.

The decade of the 1980s is certain to see an exciting new array of homebuilt designs appear due to the availability of advanced composite materials that possess characteristics of low weight, high strength and stiffness, in many cases superior to metals for sophisticated aircraft use. A whole new class of reinforcement fibers such as tungsten substrate, boron, rayon, graphite, and carbon filaments has been developed, with more than half a billion dollars already spent on this advanced technology, which may permit ultra-lightweight construction of radical designs yet unproven.

The field of plastics, the first resins to come into popular usage with homebuilders, were the polyesters, sometimes called boat resins because they were widely used in constructing boat hulls. It may be clear or bluish in color and has a short pot life—it will harden within 30 minutes after adding the hardener catalyst.

Polyesters thus have been popular with many builders of amphibian aircraft because it makes easier the job of laying up the hull with compound curves. Fiberglassing hulls also is popular, but you'll also find many homebuilt amphibians and seaplanes with hulls of plywood or metal.

In another chapter we'll explore the amphibian aircraft more fully, because there's growing interest in these versatile craft that can land on water or a hard surfaced runway with equal ease. If you live near a lake or river or other body of water, you may want to give special attention to selecting a water bird for your homebuilt.

Equally fascinating are the related machines referred to as hovercraft, sometimes called ground effect machines, or in Navy parlance, SES's—Surface Effect Ships. Today there are hundreds of

Bob Truax built this backyard space ship called ARRIBA (Spanish for UP!) to fly 90 miles out into space and parachute back.

"hover lovers" as they call themselves, mostly belonging to a fraternity called Hoverclub of America, Inc., Box 234, Uniontown, Ohio 44635. Again, we'll tell you most about SEV's (Surface Effect Vehicles) in a later chapter. These crazy craft can skim along over swamps, pepple beaches, and backwater shallows with the greatest of ease.

Not to be overlooked are the amazingly agile homebuilt rotorcraft, both helicopters and gyrocopters, that can operate from a dime-sized landing pad in your own backyard, fly straight up, forward, backward, or sideways, and come down with complete safety in a maneuver known as autorotation. When the engine quits in a conventional rigid-wing aircraft, you've got to keep up forward speed or stall and fall; in a rotorcraft, you can settle almost vertically, and at the last moment flare, converting kinetic energy into a soft touchdown. We'll tell you most about these fun machines, too, later on.

Like racing? The fierce competition of flashing around a closed pylon course, wingtip to wingtip with a rival speedster? Being the envy of the other guys as pretty girls gather around when you've won the big one? Then maybe you'd like to consider building a Formula One homebuilt, or a racing biplane, an unlimited class monster, or a North American T-6 Texan! There's even a little Formula-V racing class for Volkswagen-powered hotrods of 60-65 horsepower.

That just about covers the homebuilt movement, with one exception—how about a homebuilt spaceship? Don't laugh—a fellow named Bob Truax actually has built one, and with a little luck and some solid financing, he hopes to blast off, climb 90 miles into space, and then parachute back to Earth. Just goes to show the kind of imagination homebuilders today really have!

Chapter 7

On Wings Of Imagination

But they that wait upon the Lord shall renew their strength; they shall mount up with wings as eagles.

—Isaiah XL. 31

Throughout history man has tried repeatedly to grasp the secret of flight, to understand the mystery of how a simple wing can transform invisible air into a sustaining force through some transmutation of motion into kinetic energy.

Initially steeped in religious mysticism, the study of how a wing works became a science in the mind of 16th Century Leonardo da Vinci, who wrote: "To attain to the true science of the movement of birds in the air, it is necessary to give first the science of the winds, which we will establish by means of the movements of water."

Da Vinci in fact was aware of the analogy of wind and water and thus had a grasp of fluid dynamics as the real basis of flight, but not until the end of the 19th century and the beginning of the next did scientists like Nikolai F. Zhukovski, M. Wilhelm Kutta, Frederick W. Lanchester, and Ludwig Prandtl begin to understand what was really happening. Prandtl's mathematical formulation of the circulation theory of lift has been credited as one of the most useful contributions ever made in the study of aerodynamics. It was he who first recognized that an aircraft's total resistance is made up of three parts—skin friction (viscosity), form or pressure drag, and drag produced by lift—induced drag.

Today, we see many innovative attempts to control induced drag by altering the flow pattern of wingtip vortices with flat plates, drooping tips and winglets, and many homebuilders are finding ways to increase rate of climb, range, and other performance parameters with such devices.

At any gathering of homebuilt enthusiasts today you'll likely see a number of different kinds of airflow control devices installed on wings of various sizes and shapes—the long, thin, tapering wings of sleek sailplanes, the squarish wings of pylon polishers, elliptical wings made famous by the RAF Spitfire of World War II, delta wings reminiscent of SSTs and the SAAB Viggen short-field fighter and, more recently, the "Star Wars" futuristic swept wing of Burt Rutan's canard type "glass-backwards" VariEze homebuilts.

Which one would you choose for your homebuilt? Aesthetics aside, what kind of performance are you after—a high cruising speed, high lift, fast climb rate, low stall speed, or a combination of any of these characteristics? And what about strength? Will you be pulling 6 or 7 Gs in aerobatic maneuvers, or only one G floating along in a sailplane?

Rounded wingtips may look appealing to the eye from an artistic point of view, but they may cost you a pretty loss in climb capability, unless you square off the ends to gain the maximum length of the wing's trailing edge. The concept that a wing's effective span is based on the span of its trailing edge was first postulated by the German aerodynamicist Sighard F. Hoerner, and others in the 1920s, who first grasped what is called the momentum theory of lift—the wing simply deflects the air downward as it flies, producing the reactive force we call lift.

Flying at your airplane's best rate of climb airspeed (V_y) only half your engine's thrust horsepower is utilized in producing lift, and the other half goes into propulsion. The faster you fly, the less engine power is consumed in producing lift and the more is used up in producing forward flight, as the drag increases.

This point was not clearly understood at the turn of the century. Inventors such as Simon Lake, who first conceived of the idea for a ballasted, even-keel submarine and adapted it to a concept for a dirigible powered with a tilting rotor, posed an idea that has been born again recently in an effort to develop VTOL dirigibles with good cruise speed. To Lake, the whole concept of heavier-than-air flight was ridiculously inefficient at best.

When I interviewed the old inventor, some years ago, he told me: "Now, as one horsepower is supposed to be capable of lifting 33,000 pounds at a rate of one foot per minute, it is evident that an

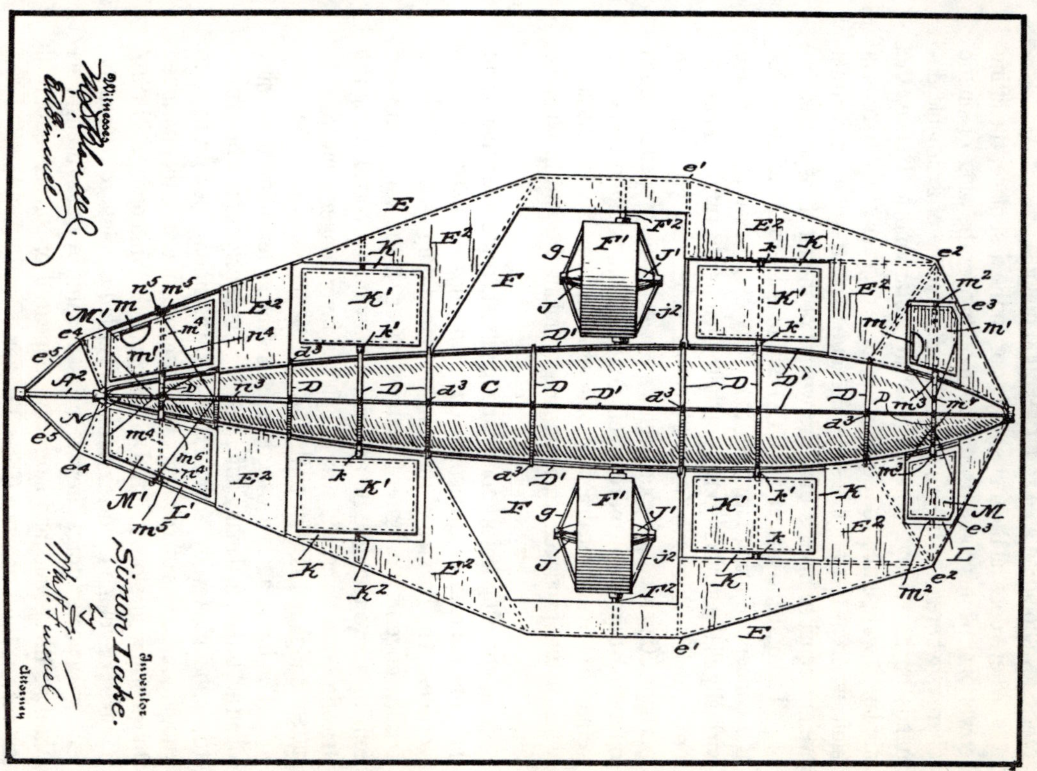

Submarine inventor Simon Lake patented a dirigible air ship in 1909 with modern-type tilting rotors for VTOL performance.

airplane's efficiency is only a very small fraction of one percent. With one percent efficiency, the flying machine should be able to lift 330 lb/hp and hover with that load. And with ten percent efficiency, it should lift 3300 lb/hp!"

Today, of course, we have a better understanding of the laws of aerodynamics, but nowhere near the efficiency envisioned by Lake has ever been achieved. The wooden Turner T-40, for example, weighs 1650 pounds gross and with a 150 horsepower Lycoming flat-four engine lifts only 11 lb/hp. According to Lake, with one percent efficiency, a 150-horsepower engine working at ten percent efficiency should be able to lift 495,000 pounds!

Where does all the lost energy go? In the first place, the lifting power of airplanes comes from air in motion over the wing, or kinetic energy, whereas in Lake's mind the lifting force of a dirigible depends on its relative bouyancy to the air around it. Hence, the airplane's weight must be equal to the force of gravity for it to remain aloft, whereas the dirigible stays aloft due to neutral density, its total weight equal to or less than the weight of the air it displaces.

Today, lighter-than-air (LTA) enthusiasts are all excited over the concept of using dirigibles for heavy cargo carrying, because virtually all the available power of their engines goes into propulsion, not into lift. But with heavier-than-air machines, you first lose perhaps 20 percent of your engine's power to the propeller, which operates at around 80 percent efficiency. Then, additional power is required to overcome drag to produce forward velocity (propulsion), which in turn produces lift, by pushing or pulling the wing through the air.

In choosing a wing for your homebuilt, remember that about half the induced drag occurs at the wingtips, in the form of tip vortices. The balance of the wing's induced drag occurs along the trailing edge in the form of bound vortices. If the wingspan were infinite, there could be no tip vortex, only bound vortex along the trailing edge. Which is why sailplanes favor longer wingspans to attain a higher L/D (lift to drag ratio).

The battle against wingtip vortex drag is becoming something of an art today, and many different ideas and designs have been used to combat it. Aerodynamicist Hoerner was among the first to go to modified wingtips of a conical camber shape, similar to that assumed by pelicans in slow flight. Dr. Richard T. Whitcomb of NASA/Langley developed "vortex diffusers" called winglets to reduce the energy of vortex drag. They were first adopted by Burt Rutan for his VariEze homebuilt, and later have appeared on such commercial aircraft as the Longhorn Gates Learjet.

Witold Kasper, of Bellevue, Washington, developed a remarkable high-lift, slow-flight wing covered by patents. His research craft, a tailless, swept-wing configuration, uses a reflex profile, the NACA 8-H-12 developed initially for helicopter blades, with wingtip washout produced by a triangular stabilizer located at the trailing edge of each wingtip, adjustable in flight from the cockpit.

To offset spanwise flow of the boundary layer and the tip stall characteristics of swept wings, Kasper added flow fences at each wingtip, the aft end of which deflects outward to serve as rudders. Each rudder carries a large aerodynamic balance which, deflected inward, creates a spoiler effect that decreases lift and induces roll. Deflected outward simultaneously, the rudders together act as air brakes and spoilers. Kasper also added patented automatic tabs at the trailing edge of each elevon surface, to avoid adverse yaw when elevons are used as ailerons.

A main feature of the Witold Kasper design are a large leading edge flap and split trailing-edge flaps that open together, developing a lift-induced vortex flow above the wing at speeds below a normal stall. He claims this increases the wing's useful load capacity from 60% to 100% of empty weight by eliminating the conventional tail, and decreases the wing's total drag by some 30%, correspondingly increasing cruise speed and range.

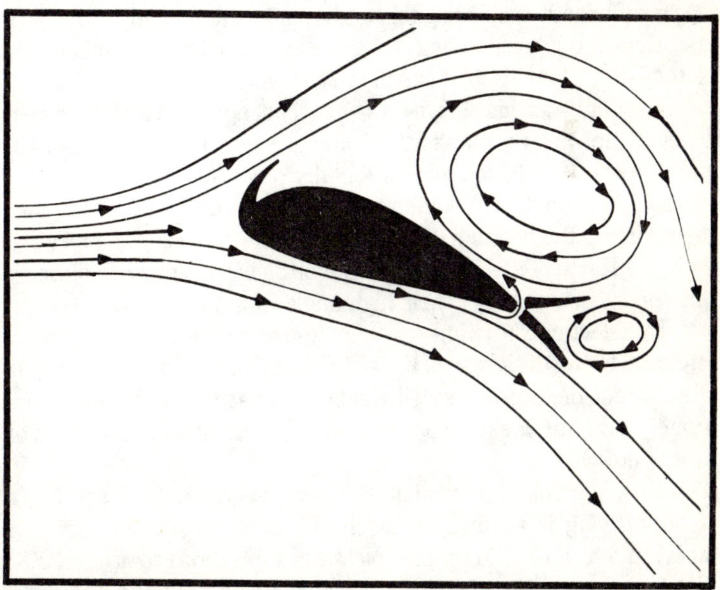

Witold Kasper designed low-speed airfoil with large leading edge flap and split trailing edge flaps to develop vortex flow over top of wing.

End plates of Ken Striplin's modified Kasper Wing ultralight.

Ken Striplin, who obtained permission to build an experimental ultralight aircraft using the Kasper design, concluded his machine was insufficiently stable in pitch, and abandoned the effort in favor of an original powered flying wing design employing similar end plates—the FLAC Foot-Launched Air Cycle.

Kasper based his design on a novel concept that vortex-induced lift over the wing permits birds to achieve zero-speed landings. He came to this conclusion after a lengthy study of bird flight, and it explained to him the phenomenon that when birds flare for a landing, their wingtip feathers spread out in a manner that causes a forward flow. To Kasper, this meant that the birds were simply reinforcing the vortex generated in their high angle of attack flareout.

In selecting the wing for your homebuilt, give some special attention to the thickness of the airfoil section, to allow use of a spar of sufficient depth for desired rigidity, and at the same time have proper depth for use of a specific airfoil curve suitable to the kind of flying you want.

NACA studies show that the best average thickness for a standard wing is about 12% of chord, unless you're designing a racer, in which case keep the thickness ratio down to around 9%. Wings that are tapered and cantilevered require a depth of around 15%.

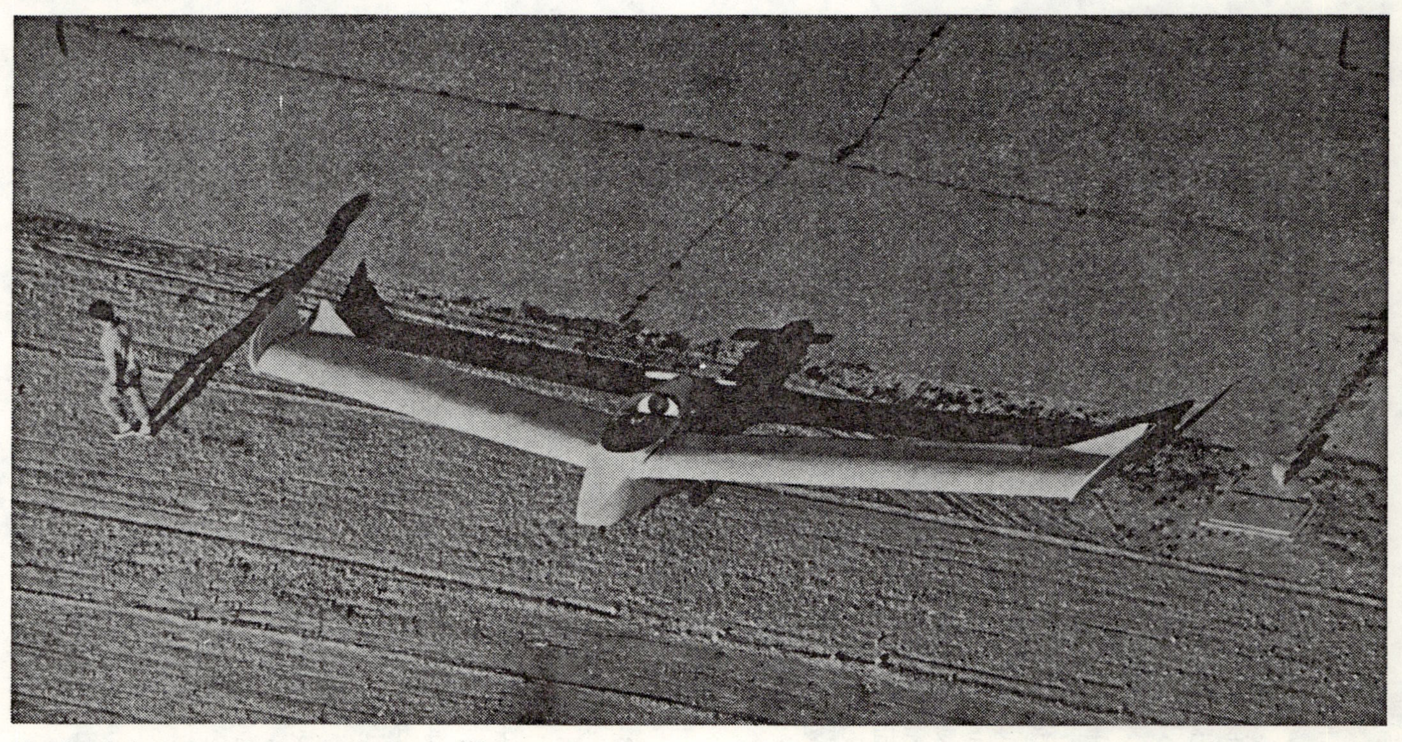

Ken Striplin built ultralight aircraft with Kasper Wing.

While some homebuilders simply "eyeball" their wing curves, it's best to take advantage of the millions of dollars NACA and NASA have spent on computerized airfoil design programs, proven out in full-scale wind tunnels. About a dozen standard airfoils are in use today, including Colonel Virginius E. Clark's Y section; the "USA" series developed by the United States Army; the "N" series from the Philadelphia Navy Yard; the "M" series developed by Dr. Max M. Munk at NACA; and still more developed at NACA research centers in the 1920s and 1930s.

A few foreign airfoils are still in use, including Gustav Eiffel's series, the British RAF curves, and the time-tested Gottingen shapes from Germany. More recently, in this country, NASA researchers have turned their attention to the development of exciting new airfoils for General Aviation and for homebuilders, a spinoff from modern aerospace technological development efforts. Elaborate digital computer programs and fluid flow theories are used, unavailable to earlier researchers of a decade ago who developed our laminar-flow wings—airfoils that perform exceptionally well under laboratory conditions but degrade severely in actual use, when the smallest particle of dirt can affect the flow pattern.

The first breakthrough in the new NASA series was achieved by Dr. Richard T. Whitcomb with the GA(W)-1 airfoil (GA for General Aviation and W for Whitcomb). Actual flight testing verified the outstanding performance predicted by the computer, and led to a second design, the GA(W)-2. Dr. Whitcomb, remember, had also developed the area-rule and the super-critical wing for transsonic and high subsonic cruise craft, and research on the supercritical airfoil for the ATLIT (Advanced Technology Light Twin) program led to modifications that resulted in the GA(W)-1 shape. Subsequently, NASA changed the GA(W)-1 design, by reducing its thickness from 17% to 13% though retaining its excellent camber, and so reduced its fairly high cruise drag coefficient.

Today, NASA is considering adopting a new airfoil designation system, dividing them into groups whose prefixes designate their function. Thus a low-speed airfoil becomes LS, a medium-speed airfoil MS, and a supercritical airfoil SC. The GA(W)-1 is now called the LS(1)-0417—the 1 means it's the first of a new series, 04 indicates a design lift coefficient of 0.4, and 17 is its maximum thickness in percent of chord. The GA(W)-2 airfoil offers excellent performance benefits for homebuilts, outperforming the GA(W)-1 both in high and low speed flight.

Virtually any reasonable airfoil will fly and give performance if properly built by the amateur constructor, and in fact you may be

Profile view of Ken Striplin's Kasper Wing ultralight.

Ken Knowles' Thorp T-18 uses folding wing design of Lu Sunderland for easier trailing and storage.

wasting your time going to fancy conical cambered wingtips, split flaps and slotted leading edges to improve low-speed flight. It has been said that the best low speed device ever invented is a low wing loading.

Interestingly, the unique control system of Dr. Paul MacCready's Gossamer Albatross channel-hopping MPA, and his earlier Gossamer Condor MPA, mark another breakthrough in aerodynamic design. The Condor, first MPA able to execute a figure-8 racetrack pattern successfully in the Kremer Competition, carried its stabilizer up front, canard fashion, with small ailerons linked to the stabilizer's trailing edge. To turn left, the ailerons were moved normally to achieve a left bank of the stabilizer, which in turn caused the main wing to follow. It overcame an earlier tendency of adverse yaw to pull the wing in an opposite direction, or to slip sideways into the ground.

A growing number of homebuilt aircraft are being fitted with foldable wings for road-towing and for easy storage in the home garage to save hangar rental. Lu Sunderland, of Appalachin, New York, redesigned the wings of the Thorp T-18 so that one man can quickly and easily fold and unfold the wings, which he calls "convertible." The wings are locked with a fail-safe device to make it impossible for them to fold up in flight, as happened to the late Dewey Bryan during an Oshkosh demo flight in his Bryan IV roadable aircraft.

Chapter 8
The Power Train

The day is gone when you could run down to the corner war surplus store and pick up a slightly used Pratt & Whitney radial engine, never flown faster than sound by a little old pilot in tennis shoes, for only a few bucks. It takes some pretty selective shopping today to locate any good used aircraft engine in the back of Joe's Hangar at any price. That's why homebuilders are turning more and more to converted automotive powerplants, motorcycle engines, and lower-powered snowmobile and chain saw jobs, for ultralights.

In the beginning, there was steam power, used in experimental aircraft by such men as Samuel P. Langley and Sir Hiram Maxim to get airborne. But since their introduction in the 1860s, internal combustion engines, operating on the four-stroke cycle, have been universally accepted for aircraft, though problems do occur when switching a road engine to sky use.

Early aircraft powerplants were so unreliable that a forced landing was the accepted thing on almost any cross-country flight, and pilots had to double as mechanics to make field repairs away from home. It was customary to carry along a complete set of tools, extra engine parts, and a spare propeller stuck back in the fuselage. In the 1920s you could pick up a good used Model T or Model A Ford engine from the local junkyard and stick it on an airplane backwards, so that the propeller attached to the "rear" end of the crankshaft.

The rear mains often leaked oil badly, and the bearings had a habit of coming apart, so today aircraft engines are fitted with larger front sleeve bearings or wide ball bearings to handle the higher-torque loads imposed in turning the propeller.

Torsional vibration continues to plague homebuilders converting automotive engines to aircraft use. Caused by power pulses generated by higher-rpm automotive powerplants, these vibrations must be dampened to prevent propeller failure in flight, not a rare occurrance. One dedicated homebuilder, Bud Rinker, of Santa Barbara, California, ran a Corvair Spyder six-cylinder turbocharged engine on a test stand for 50 hours; he found some torsional vibration which he traced to a reaction between propeller inertia and the

Bud Rinker ran 6-cylinder Corvair Spyder on test stand for 50 hours successfully.

compression strokes of the engine acting through gear backlash. The vibration diminished above 1500 rpm.

Because automotive engines are designed to operate at higher rpms than aircraft engines, the crankshaft rpm must be reduced by special reduction gearing to prevent the propeller from over-revving. Reduction drives have been in use in aircraft ever since the Wright brothers installed one with a 3:3.1 ratio in 1903.

Volkswagen engines are widely used in homebuilts today, and were already in use in Europe two decades earlier in homebuilts like the Jodel and Turbulent. Like other auto engines, the VW powerplants produce their rated power at high rpm, and so must be geared to match the propeller's most efficient tip velocity.

One of the first VW reduction gear units was designed by Don Stewart, an Allegheny Airlines captain and designer of a popular little homebuilt called the Headwind. He designed a reduction unit called the Maximizer, with a reduction ratio of 1.6:1, permitting a power output at the propeller of 32 instead of 18 horsepower, by permitting the crankshaft to turn at 3600 rpm, swinging the prop at 2200 rpm.

A number of firms now turn out special VW conversions for homebuilders, and one of the foremost is Joe Horvath's Revmaster, at Chino Airport in California. The main goal in such conversions is to permit an engine designed for easy cruising on a highway to handle the complexity of demands for multi-gaited operations, from starting to taxiing, full-power climbout, high-altitude cruise, power-off letdowns, and so forth.

The Revmaster R-2100 already has accumulated thousands of hours of actual flight time, a number of them equipped with Maloof two-position controllable propellers and Ray Jay turbochargers. Normal operation range is from 2600 to 3400 rpm, developing 65 horsepower at 29 inches manifold pressure and 3100 rpm, or 75 horsepower at 37 inches (turbocharged) and 3200 rpm. Takeoff power is rated at 75 horsepower, and emergency power at 85 to 90.

Horvath developed a special crankshaft with a 78mm stroke, 9mm longer than the stock shaft in a 1600 cc VW engine. The shaft carries extra counterweights to provide a flywheel effect in absorbing power impulses and counter-balancing forces that tend to produce vibration, resulting in extremely smooth operation. Revmaster uses special cast steel cylinders and aluminum pistons of both 8mm and 92mm size. Dual ignition is an option, and the intake manifold is specially designed to distribute the air and fuel mix uniformly to all cylinders throughout the speed range.

Another advocate of dual ignition for VWs is Lloyd Paynter of San Diego, California, whose experimental conversion is called the

Joe Horvath (left) makes dyna test of new Revmaster R-2100 VW-type engine, one of the finest flying today.

Hummingbird. A work of art, according to fellow EAA members, Hummingbird may soon be made available to homebuilders.

Among the bigger automotive conversions, Fred Geschwender's turbocharged 351 and 460 cubic inch Ford V-8s have been installed in a number of agricultural dusters with great success, and are finding a new market as powerplants for 2/3 scale homebuilt warplane replicas.

Joe Schubeck, of Malibu, California, has developed another V-8 engine called Stage II that can deliver up to 800 horsepower, relying on his earlier experience manufacturing high-performance racing engine parts for top-fuel dragsters, based on the famed Chrysler

Fred Geschwender pioneered in converting Ford engine to aircraft use. This is his 460 CID 430 hp fuel injected model.

Hemi (hemispherical conbustion chamber) engine of the 1950s and 1960s.

In Wichita, Kansas, Dave Blanton has converted the 2.3-liter turbocharged Ford 140 CID engine with a 2:1 reduction drive, that delivers 180 horsepower at 5400 shaft rpm and 48 inches MP. And one day soon you'll be able to pick up a used Model T3 turbocharger in auto wrecking yards—Garrett Corporation's AiResearch Industrial Division (AID) contracted with Ford Motor Company to deliver some 200,000 units for Ford's 1980 model year.

Basically, a turbocharger is a turbine-driven air compressor applied to an internal combustion engine to increase the weight flow of air introduced into the engine for combustion, with the energy required to operate the turbine obtained from its exhaust gases. Introducing compressed air into the engine combustion chambers allows a greater quantity of fuel to be introduced before the optimum air/fuel ratio is reached.

When turbocharger, engine, drive train, and vehicle are properly matched, the result can be a 20 percent increase in fuel economy; a 30 percent increase in power margin with the same fuel economy and weight; or 10 percent more vehicle weight for the same fuel economy, among several tradeoffs.

And what can turbocharging do for the homebuilder? It will permit him to cruise at higher altitudes where skin friction is less, at

Joe Schubeck's Stage II engine, a V-8, was developed from scratch, using design input from top-fuel dragster engines.

flight levels unattainable with normally-aspirated engines. The late Ken Rand's KR-II could fly close to 200 mph at 18,000 feet, with his turbocharged Revmaster engine.

Dave Blanton's Javelin Ford 140 CID conversion delivers 180 horsepower.

So, even though the big aviation engine companies are not now producing powerplants of under 100-horsepower, you can still come up with an inexpensive conversion engine that can operate with fuel economy geared to today's soaring gasoline prices.

Every now and then a brand new engine appears on the scene to help fulfill the need of the homebuilder for some kind of powerplant that can provide reliable and economical service at the right price. At press time, one new entry in this field was the RW-100 developed by RotorWay Aircraft Inc. of Tempe, Arizona, whose president, B. J. Schramm, designed the Scorpion RW-133 helicopter for homebuilders, and offered a special engine to go with it.

The RW-100, according to preliminary announcements, "is a fixed-wing powerplant but not an ordinary one." Not that the powerplant has fixed wings—it's a specially designed job for non-rotary wing homebuilts in the 100-horsepower range, direct drive, and offered in both single and dual ignition versions, as a water-cooled, not air-cooled, powerplant.

Preliminary design data included these specifications:

Horsepower	110 at 3300
Weight	175 lb. (dry)
S.F.C.	.38-.4
Cooling	Water (140°-180°F)
Oil Pressure	40-75 lb.
Oil Temperature	140°-210°F
Compression Ratio	9.6-1
C.I.D.	133

From Japan comes news that Komatsu Ltd. is planning to market an engine called the Xenoah, depending on the outcome of a feasibility study to determine whether there is sufficient demand for it. Komatsu produces a wide range of small, high-speed, two-stroke piston engines that have become widely popular for propulsion of many kinds of vehicles, and in 1978 won Type Certification from the Japanese Civil Aeronautic Bureau (J.C.A.B.) for their Model G72C-C three-cylinder aircraft engine. They then applied for U.S. FAA Type Certification, under FAR 21.29. The engine reportedly can deliver 60 horsepower on takeoff at 6080 rpm, or 54-hp max continuous at the same rpm. Cruise power is 45 hp at 5530 rpm. The engine burns 100/130 octane avgas at 0.92 lb/hp/hr. Displacement is 44.34 CID, compression ratio 7:1. The engine weight is 156.95 pounds dry.

Japanese Xenoah 3-cylinder engine delivers 60 horspower.

Molt Taylor developed an aircraft conversion of the Kawasaki 1000Z motorcycle powerplant.

Molt Taylor, designer of the Taylor Aerocar, Coot amphibian, and Mini-IMP, has run extensive studies of ways to modify the Kawasaki 1000Z motorcycle engine for aircraft use. Taylor approached the problem carefully and came up with an entirely new chain-drive reduction unit. Says Taylor: "It is our opinion that any sort of elastic (torsional) drive system between a piston type engine and a high inertia item like an airplane propeller must have some sort of torsional damping included to be successful. These dry fluid drive systems can be very simple: they are light weight and cost very little. They have the further advantage of reducing the torsional excitation of the relatively light weight crankshafts found in high production engines such as automobile and motorcycle engines. These were never designed for such an application as driving a high inertia element like an airplane propeller."

Taylor developed a Flexidyne drive shaft unit for his Mini-IMP, an adaptation of the Dodge Flexidyne Dry Fluid Drive which won FAA certification for use in his flying automobiles. He also considered chaining together two 1000Z Kawasaki engines to form a "unitwin" powerplant with overrunning clutches, driving a single shaft and propeller.

That's part of the fun of the homebuilt movement—see a challenge and go after a solution, without being fettered by tradition!

Chapter 9
You Can Fly On One Wing!

Should a homebuilt aircraft, designed for fun flying, have one wing or two wings? Biplane afficionados insist there's nothing more romantic than having a double set of wings, held together with singing wires, while monoplane freaks point out that if God had intended flying critters to have two sets of wings He'd have created them that way.

Of course, there are a bunch of flying insects with multiple sets of wings, if you want to be picky, but in bird country, no way. Not even a laid-back chicken hawk. Or even one that isn't chicken. And don't forget that long before the Wrights flew a bipe at Kitty Hawk, William Henson and John Stringfellow's steam-powered model planes were flying on one wing!

Despite a few early birds like Louis Bleriot's Model XI Channel Hopper and Alberto Santos-Dumont's delightful Demoiselle monoplanes, the influence of the Wrights, right or wrong, lasted nearly a quarter of a century, and the combat ships of World War I were two-wingers, and sometimes three.

Between the two world wars most private sport planes were double-deckers, and the Wacos, Travel Airs, Great Lakes, Stearmans, Gipsy Moths, etc. more or less ruled the skies. But even then graceful monos were appearing on the airways—Fairchild, Ryan, Lockheed, Stinson et al were carving a new style, to say nothing of National Airways' hot Air-King Four high-winger that could carry a pilot and three passengers in comfort.

And so it came to pass that biplanes began to fade away with the coming of the pretty little parasol homebuilts of the 1920s and

1930s, as recounted in earlier chapters—the Corben Baby Ace, Pietenpol Air Camper, and Heath Parasol really got the homebuilt movement moving. While many of these prewar lovelies are still flying, much of the postwar boom in amateur aircraft construction is due to Peter M. Bowers of Seattle, Washington, whose Fly Baby, a low-wing wooden sportster, took to the skies in 1960 and two years later won the EAA design contest as the best homebuilt flying.

Fly Babys actually are convertible to biplanes and/or floatplanes, but the majority are easy-to-build, easy-to-fly monoplanes, and well over one hundred are now flying. Fly Baby uses push-rod aileron controls because each wing system is separate and easy to disconnect for folding back prior to road-towing. Standard aircraft components may be used in Fly Baby construction, such as engine mounts, cowlings, fuel tanks, and undercarriages, and it will still be officially an Amateur-Built craft providing you do more than 50% of the work.

The Federal Aviation Administration no longer has the jitters over homebuilts, incidentally, largely because of the excellent reputation established over the years by members of the Experimental Aircraft Association. The safety record of homebuilts is, in fact, equal to if not greater than that of General Aviation factory-built ships.

In 1978, FAA Administrator Langhorne Bond visited the EAA Fly-In at Oshkosh and complimented the members on their good performance, and the following March the FAA issued a Notice of Proposed Rule Making (NPRM) to make licensing of exhibition, air-racing, and amateur-built aircraft easier. Rather than requiring an annual certification inspection by the FAA, the proposal would extend the certification indefinitely and allow annual inspections to be done by the homebuilder.

The inspection could be conducted by the qualified original builder of the aircraft, or (by) a certificated mechanic holding an Aircraft & Powerplant rating," the FAA declared. "At the time the aircraft's airworthiness certificate is obtained, the builder can obtain a Repairman Certificate which will allow him to perform required inspections on that aircraft."

The proposed inspection system would not only take much of the workload off the shoulders of FAA inspectors, "it would further reduce the burden on owner/builders of such aircraft by allowing them to utilize their specialized skills to perform annual inspections of aircraft they have built," said the FAA. "For non-builder owner/operators, the cost of obtaining annual inspections would be more than offset by relief from the annual recertification requirement and by the added flexibility made available for selecting the time and

Peter M. Bowers' Fly Baby is a top homebuilt monoplane, easy to build and fly, can be converted to a biplane.

place of the annual inspection." (The proposal has since become law).

In short, the homebuilt movement has come of age, and the government has expressed recognition of this fact and is cutting red tape as much as possible, consistent with insuring safety in building and flying experimental aircraft.

Where Fly Baby can cruise along at 110 mph behind an 85-horsepower engine, climb at 1000 fpm and range 320 miles at an empty weight of only 605 pounds, a still smaller monoplane, the Evans VP-2, gets along quite well behind a 1600 cc VW conversion. The VP-2 weighs 440 pounds empty and cruises at 75 mph, with a climb rate of 600 fpm, and is among the easiest experimental monos to build.

Bud Evans was after ultimate simplicity in design and construction in his VP-2 and its single-place sister ship the VP-1. The former Convair design engineer went all out and hung the wings on a pair of 12-foot spruce spars that are simply straight boards. The VP fuselage is built around three bulkheads and four longerons covered with a flat plywood skin. The ailerons hang onto the rear spar with a couple of eyebolt hinges, and while they feel as heavy as ironing boards, they work.

Another small VW-powered homebuilt is the Headwind, which came out of the imaginative head of an Allegheny Airlines pilot named Don Stewart, who was inspired by the early Aeronca C-2 Flying Bathtub. Stewart later modified the design and called it Headwind B, with easy-to-make wooden ribs and a high lift NACA 4412 airfoil. The 36-horsepower VW engine swings the prop through a reduction unit he designed called the Stewart Maximizer, which can be used on other homebuilts to slow down the propeller rpm and still let the VW engine run at its more efficient higher rpm.

One of the sharpest monos flying today is Bill Statler's Firefly, a two-place tandem low winger that essentially is an outgrowth of two earlier Statler homebuilt designs, the Formula One racers Little Mike and Skeeter. A straightforward design of metal construction, it carries the Continental 0-200 engine, weighs 835 pounds empty, can do standard aerobatic maneuvers, cruise at 152 mph, and hit 160 at full power; not bad for the horsepower available! A construction manual and plans are available.

Among the smaller metal monos is Cal Parker's 310-pound Teenie Two, an improved version of his original Jeanie's Teenie, now called Teenie One. Parker's design goal was to create an easy-to-build metal aircraft compatible with the VW engine. The only special tool required is a pop-riveter that can handle cadmium-plated steel pop rivets used for most of the primary assembly.

VW-powered Evans VP-2 can carry two people, is easy to build and fly.

Teenie Two first flew in 1969, the prototype costing only $650 to build in six months' time. The mono is stressed for aerobatics but the fuel and oil systems are not suitable for inverted flight. A two-seater named Double Teenie was reported under construction at this writing. Max cruise speed is 110 mph, ROC 800 fpm, ceiling 15,000 feet, range 400 miles.

The late Ken Rand of Huntington Beach, California, was among the first to introduce composite material construction in the homebuilt field with his KR-1, which first flew in February, 1972, and the two-place KR-2, which first flew in July, 1974. More than 200 sets of plans have been sold for the KR-2 and some 6000 sets of plans for the KR-1 are out.

Rand's KR-2 can accept VW engines from 1600 to 2200 cc, and the prototype carried a Revmaster R-2100D engine with a Ray Jay turbocharger, permitting a cruise close to 200 mph at 18,000 feet. Rand was killed in a crash of the KR-2 in January, 1979 when trapped in solid IFR weather and, bucking strong headwinds, exhausted his fuel supply north of Los Angeles. His brother Dan continues the business.

The KR-1 turned in an excellent performance at the 1973 EAA Pazmany Efficiency Contest, finishing in second place with an outstanding speed range from 46.8 mph in slow flight mode to 140.8 mph in a speed run. The main gear is retractable, which helped.

First place winner of the 1973 Pazmany Contest was another excellent monoplane, Richard VanGrunsven's RV-3, an all-metal high-speed, aerobatic single-place aircraft that offers ease of construction, good range (600 miles), fuel economy (26 mpg), good payload (355 pounds), and excellent short-field capability. VanGrunsven flies regularly from a 670-foot private airstrip and uses only half the runway.

This little tail-dragger is powered with the Lycoming 0-290G of 125 horsepower, and she's so clean she'll cruise at 170 mph and top out at 195 mph. I found the RV-3 easy to fly, responsive and fast in a roll (170 degrees per second), and directionally stable on landing.

For sheer fun flying, the RLU-1 Breezy takes the prize—you sit out front in the open with nothing behind you but a wing, an engine and prop, and tail feathers hung onto an open framework. It was dreamed up by three corporate pilots whose initials form RLU (Charley Roloff, Bob Liposky, and Carl Unger, of Chicago).

Their prototype Breezy was built around a wing from a wrecked Piper PA-12 Super Cruiser, and with a 90-horsepower engine RLU-1 cruises along at 75 mph, fast enough with the wind in your face. More than 500 sets of plans have been sold, and many are flying

Don Stewart's Model B Headwind has 36-hp VW engine.

with variations in the original design. One uses the wing, tail unit, wheels and fairings, wheel brakes, and seats from a Cessna 172 and carries a Continental 0-300-D engine of 145 horsepower.

Back in the 1930s a teenager in Portland, Oregon, who built model airplanes as a hobby, fell in love with a parasol plane designed by Les Long and called the Longster. Long later was killed flying another ship, a midwinger, but the boy never forgot the Longster. Later he became a Navy Commander in charge of the Boston Naval Shipyard, and when he retired in 1970 he designed and built a little two-place parasol reminiscent of the Longster, which he called Acey Deucy.

He'd already built another parasol plane in 1960, a Corben Baby Ace D, and flew it all over New England, but wanted something he could take friends along in on his journeys. Acey Deucy has a 32-foot span and 155 square feet of wing area, the airfoil a NACA 4412. Wooden ribs with full chord gussets were used, the spruce spars braced with cross-wires. Ailerons were of the Friese type, balanced. A welded steel tube fuselage was used and the stabilizer could be trimmed with a screw jack, like in a J-3 Cub. Up front, he hung on a Continental A-65 engine and a 73 × 43 propeller.

When Acey Deucy was built he called on members of EAA Chapter 106 in Boston to give it a complete inspection, and the first flight was made from nearby Taunton Municipal Airport. She got off the ground in 200 feet, climbed like a homesick angel, and cruised at 93 mph. Others who saw Acey Deucy at EAA Fly-Ins were turned on, and Cmdr. Powell eventually drew up plans for her, and close to 100 have been sold. During the cold winter of 1978-79 he added a canopy that covered both cockpits, and picked up an extra 8 mph at the same time.

While there's not room here to tell about all the fine monoplane homebuilts flying today, let's look in on a few more, like Ken Flaglor's Scooter from Northbrook, Illinois. Ken stumbled onto a little 18-horsepower engine that came from a Cushman Scooter, and decided to fit it to a plane called the Flitnick, designed by Joe Kirk. Together they became the Flaglor Scooter, which turned out to be very responsive, yet docile and forgiving. Scooter is of all-wood construction except for motor mount, center section and fittings, and is as easy to build as it is to fly. The wing is exceptionally clean, making for good performance.

Up in Forest Grove, Oregon, Bill Johnson designed the Mini-Coupe monoplane, that closely resembled the Ercoupe design of Fred E. Weick, tricycle gear, twin rudders, et al. The wing is a modified Clark Y with good stall characteristics, and with a 1600-cc

Bill Statler's Firefly is a recent popular homebuilt monoplane with a Rolls Royce Continental engine.

VW engine it cruises nicely at 90 mph. You can build Mini-Coupe in about 350 hours; aluminum, pop-riveted.

Many European homebuilt designs are monoplanes, including the sleek, two-place tandem M.J.5 Sirocco, designed by the talented Frenchman Marcel Jurca, as a potential training and touring development of his popular Tempete. Fully aerobatic, it has retractable gear and a Lycoming 0-320 engine of 150 horsepower. Some 40 are flying the world over.

The Jodel D.9 first appeared in France in 1948 as the Bebe single-seater with a 25-horsepower Poinsard engine, and later as the Bebe D.92 with a modified VW engine. French authorities were so pleased they ordered two prototypes of a two-seater version, the Jodel D.11, with a 45-horsepower Salmson engine, and the Model D.111 with a 75-horsepower Minie engine. The D.11 was the basic model for a series of both amateur-built and commercial Jodel two-seaters. Another version, the D.119, with a 90-horsepower Continental, was approved for amateur construction.

Dick Wagner, president of Wag-Aero, an aircraft supply house in Lyons, Wisconsin, and an airline captain, felt the time-honored Piper J-3 should be born again as a homebuilt and developed a kit plane replica called the CUBy, which can take any engine from 65 to 125 horsepower, and a sister-ship, the CUBy Acro Trainer, with shorter wings. Subsequently, Wagner brought out a homebuilt replica of the Piper PA -15/17 Vagabond and named it the Wagabond, a two-place, side-by-side cross country camper than has a folding seat that makes up into sleeping quarters.

Another pretty parasol plane, the EAA Pober P-9 Pixie, was designed by EAA President Paul H. Poberezny and first flew in mid-1975. With a 60-horsepower Limbach VW-type engine it cruises at 85 mph and can climb at 500 feet per minute.

In 1948, the late David Long, Piper Aircraft's chief engineer, designed and built a hot little low-winger, the Midget Mustang, and flew it in the National Air Races that year. Robert Bushby later took over the project and developed two new versions, the MM-1-85, powered with an 85-horsepower Continental, and the MM-1-125 with the 135-horsepower Lycoming 0-290-D2 engine. More than 800 Midget Mustangs are under construction today.

In 1966, a two-place, side-by-side derivative, Mustang 2, designed by Bushby, first flew, and is available with either fixed or retractable gear. More than 700 are being built and some 50 now flying.

At Charlottesville, Virginia, the old Mooney Mite Aircraft Company has been reorganized and offers plans for a modernized

Ken Rand's KR-1 is popular VW-powered mono with retractable landing gear Some 6000 sets of plans were sold

Ken Rand's KR-2 had turbocharged Revmaster R-2100 engine, could hit 200 mph at 18,000 feet.

M-18 Mite, which Al Mooney designed back in 1947, capable of flying at 120 mph with a Continental A65-8, almost 2 miles per horsepower. The new Mite does better with a redline speed of 143 mph, 114-mph cruise, 43-mph stall, 600-mile range, with a Lycoming 0-145-B2. Gear is manually retractable; it gets off the ground in 300 feet, and gets 40 miles per gallon.

Queen of the homebuilt monoplane fleet today is John Thorp's T-18, which over the years has consistently stood as an outstanding example of design excellence and top performance. Called the Tiger, the T-18 can be built for about half the cost of anything comparable in factory planes. It's a two-place, side-by-side, all-metal, low-wing retractable monoplane that can hit 200 mph with a Lycoming 0-360 engine of 200 hp.

Thorp, who also designed the Piper Cherokee, Wing Derringer, Lockheed Little Dipper and the Sky Skooter, had max performance in mind when he drew up plans for his 18th design, the Tiger. First to fly, in 1964, was built by Bill Warwick. Hundreds of sets of plans have been sold and some 100 T-18s are flying today.

Designed to take engines from 108 to 200 horsepower, the T-18 carries the mark of genius in Thorp's flying tail, streamlined profile,

Ex-Marine fighter pilot Pappy Boyington here flies a Van's RV-3 in formation with two others. He decided to build one for himself.

Flying's a breeze in the RLU-1 Breezy!

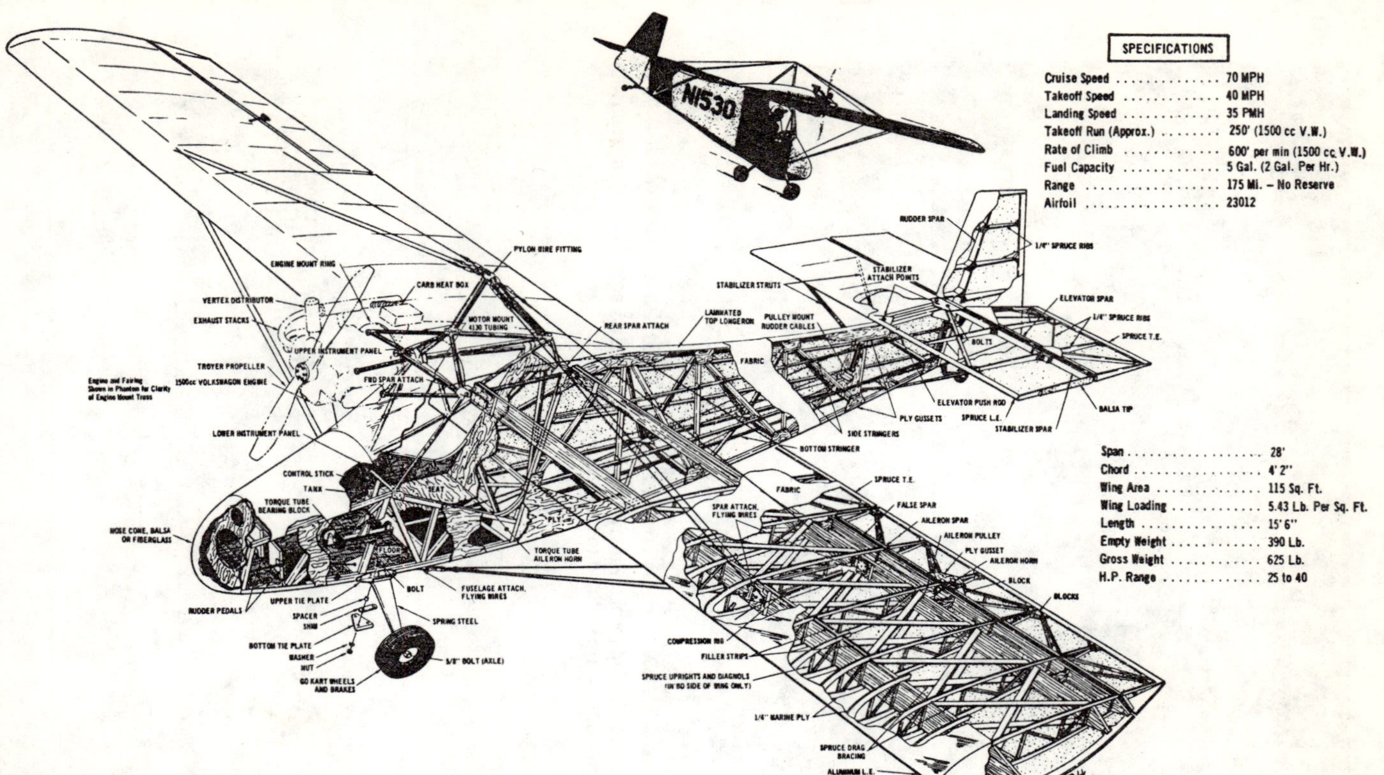

Flaglor Scooter cruises at 70 mph, is fun to fly.

Bill Johnson's Mini Coupe is a fun metal homebuilt with twin rudders.

Marcel Jurca of Paris, France, designed the Sirocco.

and "bent" wings with 8-degree dihedral in the outer panels. With Thorp's blessing, Lu Sunderland, of Apalachin, New York, designed a modified T-18 known as Model C with convertible wings, that can easily be folded back by one person for road-towing or storage in a small garage.

While this redesign was progressing, Ken Knowles, of Norco, California, built a prototype wing for his T-18 and test-flew it in 1974,

Pober Pixie was designed by EAA President Paul Poberezny.

Mustang 2 was designed by Bob Bushby.

discovering that it also had a lower flap-down landing speed. Some 200 sets of drawings for the T-18C wing have been sold, and in Australia the T-18C has won certification after rigorous testing.

John Thorp's T-18 is a top line homebuilt monoplane.

Sunderland made another design change to provide more shoulder room in the cockpit, adding 2 inches in width and lengthening the fuselage by 5 inches from canopy to fin. To improve the T-18C's stall characteristics, Sunderland also developed a modified airfoil using NASA's latest fluid flow computer programs. For a number of years, more T-18 Tigers have appeared on the flight line at the annual EAA Oshkosh Fly-In than any other homebuilt, a real tribute to the basic design by the old master, John Thorp.

Chapter 10

Two Wings Are Better Than One!

Maybe it was because Wilbur and Orville got their show on the road in 1903 with a biplane, that veteran pilots associate two-wingers with nostalgia, romance, and all the good things in sport flying.

Or maybe it was because most of the World War I fighter planes were bipes? Or that a hundred thousand cadets of World War II vintage won their wings in Stearman PT-17s?

Whatever the reason, biplanes today still invoke spine-chilling admiration at airshows, and attract attention anytime they drop into a small field for a visit on a cross-country tour. And they remain popular with the aerobatic set, for good reason—there's a greater sense of security when pulling 5 or 6 G's executing outside snaps, Lomcevaks, or other precision maneuvers, when you can look out and see something solid, held together with a maze of flying and landing wires, or at least tough interplane struts.

Not that you can't build a cantilevered-wing monoplane that can run through the whole Aresti routine in safety. Biplanes simply have a stronger place in the hearts of active airmen who enjoy the rugged beauty of a two-winger, listening to the sound and fury of wind howling through the wires.

I had my fling with biplanes in World War II as a Royal Air Force flight instructor at Falcon Field, Arizona, teaching aerobatics, instrument flying, night flying, etc., in Stearman PT-17As, and came to love the brutes with a fierce passion. They'd take any punishment a heavy-handed cadet could inflict on them, and yet the Stearman primary trainer possessed a certain sensitivity and responsiveness you just had to admire.

Today, there are a few beautifully restored Stearman PT-17s still flying, and a good number are doing duty as crop-dusters with 450-horsepower radials replacing the original 220-horsepower Continentals. But by and large, the sport biplanes you see at any homebuilder's fly-in are considerably smaller, more graceful, and in a sense easier to fly, with certain exceptions.

Consider the Pitts S-1S homebuilt version of the popular Pitts S-1 Special, designed by Curtis Pitts to replace an old Waco F he once owned in Jacksonville, Florida. In 1966 Pitts redesigned the little biplane with a four aileron symmetrical airfoil that could roll faster and fly equally well rightside up or upside down.

The Pitts had one problem—it was a bear on landings, being "short-coupled," and its treachery showed up in official NTSB accident reports. Typically, a pilot not alert on landing would let the Pitts swerve and ground loop, or use too much brakes and flip over, but proficient pilots found it a joy to fly.

The U.S. Aerobatic team in 1972 whipped all competition in France flying Pitts Specials, and Professor Art Scholl switched from his monoplane Super Chipmunk to a Pitts Special because he found it easier to stay within the performance box during competition routines.

More recently, the EAA's aerobatic trio known as the Red Devils, Tom Poberezny, Gene Soucy, and Charlie Hillard, flew Pitts Specials exclusively, until Soucy switched to Frank Christensen's amazing Eagle homebuilt for competition flying in 1978, and liked it so much the Red Devils all went to Eagles.

Regardless, you'll see Pitts Specials flying in airshow routines for years to come, and hundreds are flying simply as sport planes. Some 2500 sets of plans were sold before Pitts was bought out by a group in Afton, Wyoming, called Pitts Aerobatics. The early two-aileron model can be retrofitted with the four-aileron symmetrical wings, or you can start from scratch with the S-1C with a modified M-6 airfoil.

A Pitts homebuilt can be put together in about 2000 man hours, or you can build it faster from kits available from the Afton plant. A Pitts fitted with an 180 horsepower IO-360-B4A engine and Sensenich 76EM8-0-56 prop can hit 147 mph, climb at 2600 feet per minute, and roll through 360 degrees in 2 seconds. She's small—wingspan is 17′ 4″, length 15′ 6″, height 6′ 3″, and she weighs only 720 pounds empty or 1150 pounds gross.

Pitts Special is one of the best aerobatic homebuilts.

Frank Christensen admits he designed the Eagle I & II in direct competition with the Pitts Special, but there's a difference—the Pitts Special is FAA Type Certificated and available as a factory ship as well as in kit form, whereas the Eagle is strictly for homebuilders.

Christensen introduced the Eagle II first, in 1977, as a do-it-yourself assemblage of 25 separate kits that can put you in the competition action for $14,000 to $16,000 plus engine and propeller. Christensen, in fact, was a former Pitts driver who won the National Aerobatic Competitions, Advanced Category, in 1968. He also was team captain in 1972, when Charlie Hillard won the world aerobatic championship and Mary Gaffany won top honors for women pilots.

He ran a computer study of the market and decided there are maybe 2000 builder/pilots with sufficient skill and professionalism to put together a Pitts S-1D homebuilt, whereas there are thousands more "amateur constructors" who can assemble an Eagle.

He explains: "Eagle I and I-F are special-purpose single-place biplanes intended solely for unlimited class aerobatic competition (Eagle I has a constant-speed prop and Eagle I-F a fixed-pitch prop). The Eagle II and II-F are two-place sport biplanes fully capable of unlimited class aerobatic competition flying, but designed for additional pilot/passenger comfort and convenience. Both have full electrical systems. Eagle II uses a 200-horsepower engine and Eagle II-F a 260 horsepower engine with fixed-pitch prop."

Though similar in size to the Pitts, the Eagle has more cockpit room, smoother aileron control response, and a roll rate of 187 deg/sec. Eagles also have symmetrical wings for inverted flight, using the Christen inverted oil and fuel systems also designed by Christensen.

The homebuilder starts off assembling the Eagle with the aileron kit, which is inexpensive and instructive—for a few bucks you can find out if you have the manual skill to pop-rivet and other stuff. With each of the 25 kits comes a beautifully illustration, printed manual with everything numbered, pictured, and tabulated in a professional manner.

Eagle II-F, the homebuilder's sporting version, can hit 192 mph with the 260-horsepower engine, cruise at 167 mph, stall at 58 mph, climb at 2450 rpm, roll at 187 deg/sec, and range 362 miles. It has unlimited inverted flying capability and a vertical penetration of 1400 feet.

There's a score of other excellent homebuilt biplane designs to choose from, including the excellent, time-proven Stolp Starduster Too SA-300, one of a line of ships designed by Lou Stolp, and currently by his successor, Jim Osborne, at Flabob Airport, a

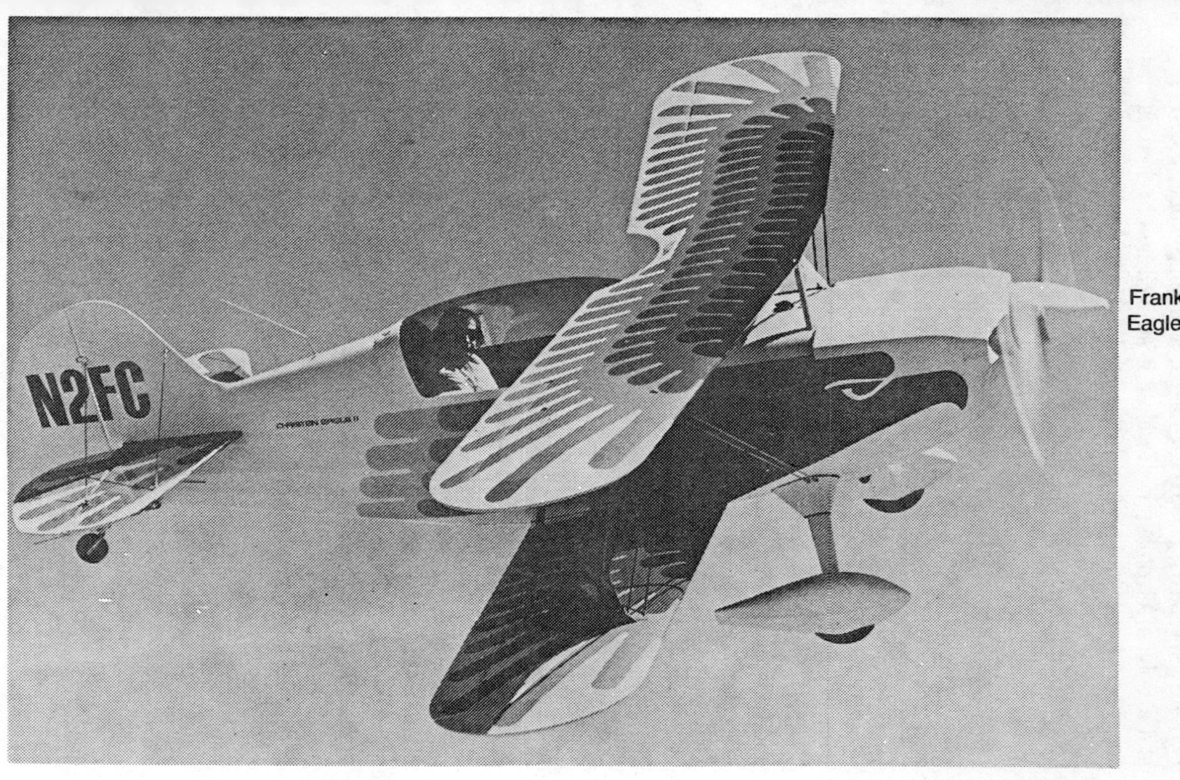

Frank Christensen's
Eagle II is a show-stopper.

homebuilder's haven at Riverside, California. Osborne first brought out a competition machine, the Acroduster Too, and followed that with the single-place aerobatic Acroduster One.

John Helton, an airline captain, took Jim's Acroduster II, N750X, back to Fond Du Lac, Wisconsin, in the summer of 1978 and won the Intermediate Category competition well ahead of 23 Pitts, two Stevens Akros, two Great Lakes, a Steen Skybolt, a CAP-10, and a Spinks Akromaster. He liked it so well he checked out his daughter, Janet, and on her 16th birthday she not only soloed the Acroduster II but also a hot-air balloon, five sailplanes, and 29 other airplanes including 18 taildraggers, seven of them homebuilts.

I had the pleasure of flying the Acroduster Too with Janet piloting and found her technique flawless. I also flew Jim's Acroduster One, and found it to be just as lovely to handle as the old Stearman PT-17As, only lighter, faster, and more maneuverable.

Lou Stolp's first homebuilt design was the single-place SA-100 Starduster that can be built for under $1800, followed by the parasol-wing monoplane Starlet. Next came the V-Star SA-900 which Osborne designed as a low-cost, low-powered sport biplane, and then Starduster Too, with more than 200 flying today.

Starduster Too, powered with the 180 or 200 horsepower Lycoming IO-360 engine, is a rugged competition ship built strongly of 4130 steel tubing and sheet stock, the wings built around spruce spars, ribs of plywood shaped to the modified M-6 airfoil section.

Starduster Too weighs about 1000 pounds empty and grosses 1750 pounds. Upper wingspan is 24 feet, and the lower span is 20 feet 6 inches for a total wing area of 162 square feet and a wing loading of 10.8 lb/sq. ft. Osborne's Acroduster Too SA750 is a smaller version of Starduster Too, designed at the suggestion of a TWA captain, Morgan Schrack, who purchased the prototype and dolled it up with a fancy canopy. Flown solo, the front cockpit is sealed and the pilot sits under his plastic shell in the rear hole in total comfort.

Acroduster One, I discovered, has a fast roll rate of 240 degrees per second and a nice cruising speed of 185 mph TAS at 2400 rpm and 24" MP. Top speed is over 200 mph. With the same elliptical wing of the Stolp Stardusters, Acroduster is even more maneuverable, the four ailerons moving in concert with a fore-and-aft stick movement. With the ailerons in neutral position, these control surfaces lie in trail with the trailing edge of the wing; pulling back on the stick raises all four ailerons slightly, to offer better roll control down through the stall.

Starduster Too is another popular aerobatic biplane.

Acroduster Too is scaled-down version of Starduster Too.

The prototype Acroduster One was demolished in May, 1976, during an air show at Corona Airport, California, when veteran aerobatic pilot Manx Kelly went far past the redline and exceeded the aircraft's design stress limits in a pullout from a vertical dive, that caused the tail to separate. Osborne has since redesigned the empennage structure to prevent such a needless accident from ever happening again, in such a fine airplane.

It was back in the 1950s that EAA's President Paul H. Poberezny headed a design team of seven members who decided to create a simple, straightforward little biplane patterned somewhat after an existing aircraft called the Gere Sport. The prototype EAA Biplane, as it was called, was built between 1957 and 1960 as a classroom project by students of St. Rita's High School in Chicago.

Powered with a Continental A65 engine, its performance was lower than expected when it first flew in June, 1960, because of improper construction of the cabane and interplane struts. It was rebuilt and a new metal prop added, engine cooling improved, and canopy installed, and when EAA took over the project they added an 85-horsepower engine. Poberezny added a bigger tail and renamed it Model P (for guess who?).

Despite its birth pangs, the EAA Biplane became quite popular and more than 700 sets of plans were sold. Many are flying today,

EAA Biplane was designed in 1950s as a school project plane.

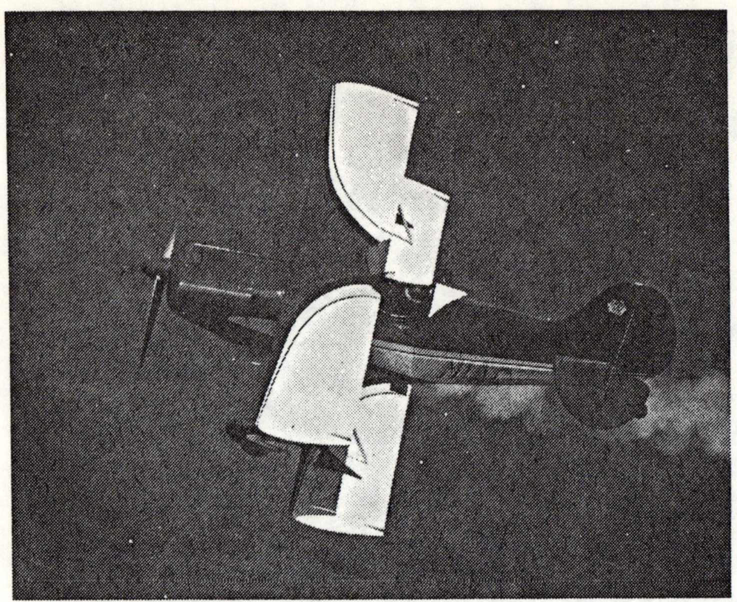

EAA Fresident Paul H. Poberezny designed the EAA Acro Sport.

and one, N177E, owned by Roy Moore of Wichita Falls, Texas, had been built by Arthur Guatreau in Georgia between 1962 and 1972 from a set of early 1952 plans. A 150 horsepower Lycoming replaced the original C-85, and a tail-heaviness vanished.

N177E's stalls became quite docile, spins were normal, and despite a slow roll rate, barrel and slow rolls were excellent, snaps fine, loops and hammerheads good. EAA Biplane's span is 20 feet, empty weight 710 pounds, and performance variable, according to powerplant used.

Subsequently Poberezny struck out on his own and designed a brand new biplane, the EAA Acro Sport, to fulfill the need for a fine little biplane with an extra good set of drawings and a construction manual to be used in school Industrial Arts programs or the Civil Air Patrol as a motivating and educational benefit.

Acro Sport has a span of nearly 20 feet and can carry engines from 100 to 200 horsepower. It was 352 days from the time Poberezny began the first drawing until the first flight, and he recalls, "within a minute after its first flight I had it on its back." The airplane has not been changed since that first flight, and more than 400 are under construction.

The prototype Acro Sport carried a 180 horsepower Lycoming engine and is an excellent aerobatic aircraft and has a fast climb of

3000 fpm. With the bigger 200-horsepower Lycoming, the Super Acro Sport climbs at 3500 fpm and has the same top speed of 180 mph. Wag Aero of Lyons, Wisconsin, supplies Acro Sport prefabricated kits, and plans are available from Acro Sport, Inc., P. O. Box 462, Hales Corners, Wisconsin 53130.

One unusual Acro Sport was built inside the walls of the Green Bay Reformatory in Wisconsin, where two inmates, Patrick Dorgan and James Mutaler, began writing letters to local flying groups seeking information on homebuilts. To their surprise Roger Davenport, president of EAA Chapter 250, drove up with a truckload of parts. With blessings of the reformatory staff, they were making good progress at last report, and their only worry, as Dorgan put it with a grin, was that somebody might break in and swipe it.

Back in 1929, a chap named Barney Oldfield went to the National Air Races in Cleveland and fell in love with the Great Lakes 2T-1 Sport Trainer biplane. When the company folded, Barney scaled the plans down to make a mini-Great Lakes, but it bombed out. So he started from scratch and built an entirely new little bipe, the Baby Lakes.

Baby Lakes was a lovely little ship, and when Oldfield died, a friend, Harvey Swack, took over the project to keep it alive. The standard Baby Lakes takes engines from 50 to 100 horsepower, cruises around 118 mph, climbs at 2000 fpm to a 17,000-foot ceiling, stalls at 50. With a span of 16′ 8″ and wing area of 86 sq/ft, her wing

The Baby Lakes is an exciting little bipe to fly.

loading is about 10 pounds per square foot, with the C-85 Continental.

Subsequently Ray Ball, of Fulton, New York, built up a Super Baby Lakes with a 115-horsepower Lycoming 0-235 and got a cruise of 135 mph TAS on 75% power and a top speed of 155. She gets off the ground in 350 feet and climbs at 2500 fpm. Barney Oldfield's concept was a design safe for low-time pilots to fly from grass strips.

LaMar Steen, a World War II Army Air Corps pilot, settled in Denver, Colorado, built a number of homebuilts and became president of EAA Chapter 43 as a sort of hobby activity. As an aerospace teacher at Denver's Manual High School, Steen's principal asked him to find some way to interest students in aeronautics as a pre-professional starter program.

LaMar got busy at the drawing board on a two-seater biplane, one the kids could go for a ride in when it was finished. The design work began in mid-1968 and construction in August, 1969. A funding project fell through, but LaMar was in love with the project and financed it himself for about $5,000.

In October, 1970, after three classes of youngsters had built the Steen Skybolt, it was ready to fly. The kids had built it all, including the welding, woodwork, engine installation, etc., and when LaMar zoomed off the runway at Jefferson County Airport they yelled their hearts out. One by one they all got rides.

Powered with a 180 horsepower Lycoming, the Skybolt, painted red, white, and blue, was a real performer . Upper and lower wings were both symmetrical for inverted flight, the top wing a NACA 0012 airfoil, the lower wing a NACA 6300 series airfoil. The ship was designed for stresses of +12 and −10 G's for full aerobatic performance.

Weighing 1080 pounds empty, Skybolt cruises at 130 mph, has a top speed of 145 mph, and 2500 fpm climb to a ceiling of 18,000 feet. It will loop from level flight with two people aboard at mile-high Denver, can execute inverted spins and multiple inside and outside snaprolls. Control pressures are light but not overly-sensitive. More than 1000 sets of plans have been sold to homebuilders.

There's hardly room here to tell about all the other fascinating homebuilt biplanes flying today, but we can list a few more of the better-known types, such as Javelin Aircraft's Wichawk, designed by Dave Blanton. Its structural geometry is similar to the World War II Stearman PT-17A primary trainer, and it's stressed for +12 and −6 G's. Wichawk first flew in 1971. With an 0-360 Lycoming, Wichawk weighs 1280 pounds empty, can hit 140 mph and climb at 1700 fpm at sea level.

LaMar Steen's Skybolt is one of the best biplane homebuilts.

Dave Blanton's Wichawk has the lines of a Stearman trainer but is smaller and more agile.

John D. Hatz, a midwesterner from Wisconsin, used to fly an old Waco 10 around the farm country, and when it began showing signs of aging he designed a homebuilt look-alike he called the Hatz CB-1 Biplane. He tried a Continental C-85 engine but found it a bit underpowered, so went to a 150 horsepower Lycoming 0-320. Another homebuilder, Dudley Kelly of Versailles, Kentucky, talked Hatz into drawing up a set of plans so he and others could build CB-1s also. Kelly used an 0-200 engine of 100 horses and flew his ship to Oshkosh in 1975. It weighs only 865 pounds empty and cruises at 90 mph, a fun sport plane with aerobatic capability, small and comfortable.

Out at Flabob Airport in Riverside, California, Ed Marquart has turned out a number of exceptionally fine homebuilt designs, including some replica aircraft. His fifth, the MA-5 Charger, was conceived in 1961, and five years later the little biplane began to take shape with a buddy, Dan Fielder, doing the major work and financing.

Charger can take two six-footers comfortably on happy aerobatic routines. Wings are of equal span and chord with a sweepback of 10 degrees. Four slotted ailerons are used. Conventional 4130 steel tubing is used in the fuselage, and the wings are assembled on spruce spars with built-up ribs. With a 125 horsepower Lycoming, Charger, which weighs 1000 pounds empty, has a top speed of 125 mph, 42 mph stall, 1100 fpm climb rate, and a range of over 400 miles.

Ed Ma:rquart's MA-5 Charger is fun to fly.

Don Stewart, an Allegheny Airlines pilot, and a friend, Tom Raybourn, another airline pilot, spent one rainy summer weekend at a fish camp in upstate New York designing a look-alike World War I fighter-type homebuilt biplane they called the JD-2 Foo Fighter (JD = Jim Dandy!). It first flew in June, 1971 with a six-cylinder Ford

John O. Isaacs scaled down a Hawker Fury to come up with this pretty little biplane. The Isaacs Fury.

Falcon motor, but later used a 125 horsepower Franklin Four. Weighing only 725 pounds, it's a mighty little funster fighter to go shooting down imaginary bogies in.

In January, 1961, John O. Isaacs, a professor at England's Southampton College of Technology, started the design of a scaled-down Hawker Fury single-seat biplane fighter of the Royal Air Force, and the Isaacs Fury first flew in 1963, powered with a 65-horsepower Mikron III engine, later replaced by a Lycoming 0-290 of 125 horsepower. This 7/10ths scale look-alike replica weighs only 720 pounds and cruises at 100 mph, stalls at 42 mph. Nicely aerobatic, she's stressed to +7 and −4.5 G's, but not for snaps. A number of other Isaacs Furys are building in the United States, England, New Zealand, and Canada.

Chapter 11
Water Wonders

America is a nation of beautiful waterways, thousands of miles of lovely coastlines, lakes, rivers, bayous, awaiting the aerial explorer who can build and fly his own water-bird. You don't have to be a millionaire to own your own air yacht. Volmer Jensen's VJ-22 Sportsman is a fine example of a low-cost two-place amphibian, powered with a C-85 Continental that swings a pusher prop mounted atop a pylon.

Jensen likes nothing better than to toss his snorkel, mask, and fins into the baggage compartment and head for some quiet bay in the Sea of Cortez for a weekend of skin-diving far from civilization. He's also explored Canada's lake country, and many of the country's major river systems, flying low and slow along the waterways, raising flocks of water birds, and always with a safe place to land in view.

Some 750 sets of VJ-22 plans have been sold and perhaps a hundred are now flying in many parts of the world. Jensen's prototype first flew in 1958 and has logged some 2000 hours of happy flying. A spinoff of his earlier VJ-21 high wing pusher landplane, the Sportsman has a rigid and corrosion-proof hull of 1/16″ and 3/32″ aircraft mahogany plywood with 1/4″ ply at the step for maximum strength. The hull is covered with fiberglass for extra protection.

At sea level, the Sportsman gets off the water in 20 seconds. A 20-gallon fuel tank under the C.G. in the hull provides roughly four hours flying time—300 miles to be safe. Empty weight is 1000 pounds, useful load 500 pounds. The wings can be assembled from a

Volmer Jensen's VJ-22 Sportsman amphibian is just the thing for exploring remote bays and rivers.

wrecked Aeronia Chief or Champion, and the main gear, swiveling tail wheel and water rudder are all manually retractable.

For the cost, you can't beat George Pereira's little two-place homebuilt amphibian called Osprey II, which he built for $4,214.14 including a used Lycoming 0-320-A2B he picked up for $1,025. Osprey II weighs only 970 pounds empty, and that's only $4.34 a pound, less than the price of a good halibut steak.

Periera is a successful businessman who ran a building supply firm into a good thing that financed a Mooney Super 21 for business flying back in 1966. But as a former military pilot (P-38s, B-17s) he wanted something more exciting and went into the homebuilt arena for it. His first effort was a two-place cross-country homebuilt something like a Thorp T-18, but the gear failed on her first test hop and she was totaled.

Periera, who lives up in the Sacramento River delta country of Northern California, next decided to build something he could use to explore the endless hidden channels and canals in that water world, and the result was a single-place flying boat with folding wings, called Osprey I. He completed it in a year and it was so successful "it just flew right off the drawing board," he says. But—it had only one seat and no landing gear to make it amphibious. In 1971, the Navy heard

Osprey II cruises at 130 mph with Lycoming 0-320 engine.

about it, and got all excited about building a fleet of them to use in mopup operations in a place called the Mekong Delta, before the United States lost its first war, in Vietnam. So back to the drawing board again.

First, Osprey II had to be a two-placer, and Periera kept it small—1000 pounds empty weight with a useful load of 560 pounds, about like the Volmer Sportsman. He stressed the wings to 4.5 G's and geared the design to the skilled amateur builder with limited tools—a band saw, table saw, drill press, grinder, gas welder, paint sprayer, hand drill, and sander, plus hammers, saws, and chisels.

The design was two years aborning, and the construction took some 1300 man hours. He designed the hull to permit the bow wave to flow back under the wing to offer pre-lift for getting the hull up on the step, and that meant tucking the wheels up into a nine-foot wide center section, with a redundant retracting mechanism.

He chose wood construction as simplest for the homebuilder, plus polyurethane foam and fiberglass as used on Osprey I. Side-by-side seating offered a wider hull for better floatation and efficient water performance, and the sloping canopy was designed to offer a smooth flow of air into the propeller disc, plus improved visibility.

The hull bottom is covered with several layers of fiberglass covering a layer of polyurethane foam, sprayed onto the varnished plywood by a roofing firm, and then cut and sanded to a smooth shape. The result is a sandwich of plywood, foam, and fiberglass that can withstand heavy impact loads on rough water landings. The floor is flat, so no bilge pump is required.

The 150-horsepower Lycoming 0-320-A2B engine gives a true cruise speed of 130 mph at 70% power, with a Ted Hendrickson 3-bladed propeller, or a TAS of 141 mph at full throttle. Periera ran high-altitude water landing tests at historic Donner Lake in the High Sierra country, at 6000 feet elevation; she got off the water easily with two aboard. Flying with Periera out of Watsonville, Calif., I noted a sea level climb of 1000 fpm, indicating 100 mph at 2650 rpm, not bad performance!

Earl Anderson, a homebuilder who also is a Boeing 747 captain, spent 9 years and $5500 to complete a nice amphibian by 1969, called the Anderson EA-1 Kingfisher. He subsequently replaced its 100-horsepower Continental 0-200 engine with a 115-horsepower Lycoming 0-235-C1 that brought the empty weight up to 1092 pounds and considerably improved its performance. With the 0-200 the Kingfisher weighted 1032 pounds, cruised at 85 mph, climbed at 500-600 fpm, and had a ceiling of 10,000 feet.

Built for two people seated side-by-side, the Kingfisher uses a standard Piper J-3 Cub wing with stabilizing tip floats of aircraft mahogany plywood covered with fiberglass. The gear is retractable. Initially designed to take bigger engines up to 140 hp, Anderson later discouraged using the larger powerplants based on experience with the lower-powered units. More than 200 sets of plans have been sold, some 100 are under construction now, and at least ten already are flying.

Among the smaller amphibians is the Catfish Special, built down in Tallahassee, Florida, by two homebuilders, Lester Durden and Bud Bauer, who wanted a simple, inexpensive water-bird to explore the back bayou country—Lake Miccosukee, the Chattahoochee River, or maybe Apalachee Bay. They got the idea for their amphibian from reading an article on a similar one designed by Al Hoffman, of Tarpon Springs, Florida. Bauer, a cabinet maker, built the hull, and Durden, an A&P mechanic, built the wings, tail, and engine mount. They installed a 100-horsepower Lycoming, called her the Hoffman X2-M and on April 3, 1969, got her into the air from Tallahassee Commercial Airport.

Catfish cost only $800 to build, and its dry weight is 900 pounds. She cruises at 75 mph, climbs at 300 fpm and can range over the swamplands of Florida's Panhandle for 3 hours, utilizing a 7-gallon upper tank and a 12-gallon bottom tank. Her hull has exceptionally clean lines with a V-shaped prow and mid-section step. Side fairings throw the spray clear.

Both elevator and rudder control cables run along the outside of the fuselage on top, connecting with the horizontal control surface

Lester Durden's Catfish Special plywood amphibian is fun to fly exploring hidden bayous.

Larkin KC-3 Skylark is a fine two-place amphibian.

atop the vertical fin through pulleys. Dual controls permit operation from either the front or rear cockpit. A tiny Confederate flag waves bravely from the pitot-mast in front of the windscreen.

In 1961, a design got underway by the Larkin Aircraft Corporation, now of Scotts Valley, California, on a pretty little two-seat monoplane, the Model KC-3 Skylark, with side-by-seating in an enclosed cockpit and twin tail booms extending rearward. Its braced wings were of light alloy, utilizing metal ribbing and skin. Fiberglass cambered wingtips were added, drooping to serve as floats. Fiberglass lower and upper fairings were added to a tubular fuselage keel, and an optional V-shaped stepped lower unit could be attached to make her truly amphibious.

First flight took place in 1972, with a modified 1600 cc VW engine of 100 horsepower, and at maximum gross weight it could climb at 550 fpm and cruise at 105 mph. Range was 525 miles. Its wing area of 114 square feet provided a wing loading of roughly 11 pounds per square foot. Conversion from landplane configuration to a seaplane with amphibious capability takes about two hours. Basic sheet-metal construction is used, and the homebuilder can order any parts that they may feel are too difficult to make. Two Skylarks were flying by 1976.

Whereas most amphibious light aircraft built in the past have relied on some sort of tip floats for lateral stability in the water, Molt Taylor's Coot design employs a "floatwing" arrangement whereby the inboard panels are used as sponsons, eliminating the drag penalty of tip floats.

The wing is sufficiently stiff for the pilot to stand on while hand-propping the pusher engine up on its pylon, and further makes it easier to inspect the powerplant. The engine/propeller thrust line is relatively low to the waterline, eliminating the high thrust disadvantages of some amphibious designs that make it difficult to maintain trim with power changes.

Coot A has a single tail and Coot B has twin vertical fins, and in both the horizontal surface is almost directly in the propeller slipstream, providing maximum elevator effectiveness to raise the nose on water takeoffs and so avoid plowing.

The Coot's general arrangement permits pilot and passenger to step aboard the hull-cockpit when operating off the ground without using special steps, by merely stepping onto the cockpit's flat floor. Both seat bottoms fold up so the occupants can stand on the floor and fish over the side in some sleepy lagoon. You also can step over the windshield onto the foredeck, which is coated with a non-skid surface, for boarding or disembarking.

The landing gear is tricycle type, but a conventional tailwheel gear may be installed optionally. The tri-gear is retracted manually, the nosewheel folding forward and the mains folding aft out of the slipstream for improved performance. No wheel well doors are required.

The wings may be folded by one person. They're attached with tapered steel pins pulled with a simple tool. A conventional boat trailer may be altered for transporting and storing the Coot. Wingtips are provided with hand grips for folding, and stainless steel spring tip skids protect the tips during beaching operations. Diagonal wing ribs make unnecessary internal wire bracing.

Taylor now provides all parts included in the materials list including the fiberglass hull shell, something that could easily take the homebuilder a full year to build. The shell incorporates a six-inch baggage space extension shown in plans for Coot A with the long hull (there are no plans to supply hulls for Coot B).

Prototypes of both Coot A and Coot B have been flying for several years, the Coot A with a Franklin Sport 6 engine (180 horsepower) and Hartzell constant-speed propeller, and the Coot B with a Continental C-125 engine and Flotorp controllable prop of pusher configuration. According to Taylor, sea-level performance is spectacular, with extremely short takeoff runs on land or water and a steep climbout. Both models have been flown at design gross weight and a sea level cruise at something over 110 mph.

The higher-powered Coot A (180 hp) is meant for higher-altitude operation from mountain lakes, permitting a gross takeoff weight operation from 6000-foot altitude.

Taylor adapted the basic Coot design concept from a family of amphibious marine assault gliders during WW2 for beach assaults. The Floatwing design has proven to be very practical, making it possible to land on the ocean and taxi through the surf.

Though Coot A was first to fly, there's little difference between the two configurations. The A has been proven easier to build and offers better performance in the air due to its light weight and single tail. The B's wings fold with the leading edges up, and the A's fold with the leading edges down, giving better road stability when towing.

Structural gross weight of the Coot is 1950 pounds if plans are followed exactly, but actual allowable gross weight in flight depends on powerplant and propeller used, and actual flight conditions.

Unquestionably, the top of the line in homebuilt amphibians remains the Spencer Amphibian Air Car Model S-12-E, a four-place composite construction pusher type that closely resembles the Republic SeaBee—the same designer was responsible for both, Percival H. Spencer of Sun Valley, California. The Air Car, considered to be the ultimate sportsman pilot's flying yacht, can be built for under $25,000, but you have a flying machine you couldn't touch commercially for under $60,000 to $70,000 today.

Spence, as he is known to everyone, and a partner, USAF Colonel Dale Anderson (Ret.), built the prototype Air Car after Andy insisted that Spence complete a set of plans he'd had on his desk for years, while inventing other money-making things like the Wham-O Bird toy. His father, Christopher Spencer, may be remembered as the man who designed the famed Spencer repeating rifle that practically won the Civil War for the North—200,000 were turned out in a manner that launched America into the era of production line manufacturing.

Whereas Spencer developed the Republic SeaBee from an earlier homebuilt design, his S-12D, the Air Car came after he refined the SeaBee design to become the Trident TR-1 Trigull-320, now certificated by Canada's Department of Transport. Back to homebuilding, he developed the Air Car as plywood version from the same family tree. Some 150 sets of plans have been sold and perhaps a dozen are now flying.

The prototype Air Car, N11DA (for Dale Anderson, who financed it), was developed from a patent granted to Spencer in 1950, and in 1970 it was powered with a 180-horsepower Lycoming and a 72-inch constant speed two-bladed Hartzell prop. Construction includes a braced wing of wood spars and plywood, fiberglassed, with tip floats. The hull is conventional with wooden bulkheads, longerons, and skin. A welded steel tube cabane structure carries the wing and engine mounts and attachment points for the landing gear.

Spence used the NACA 4415 high-lift STOL type airfoil and an all-flying tail that together provide smooth handling down to the stall. After changing engines to the 260-horsepower Lycoming 0-540 and widening the wingspan, Spence and Andy ran high-altitude water trials at Lake Tahoe on a day when the density altitude was close to 9500 feet. The takeoff runs averaged 28 seconds.

A still bigger engine was used next—the Teledyne Continental Tiara of 285 horsepower, qualifying the Air Car with a new designation—S-12E. A three-bladed Hartzell propeller was matched to the Tiara, and it can go into reverse for water-backing maneuvers.

At this writing Spencer was considering changing from ailerons to spoilerons to permit full-span flaps to beat the adverse-yaw problem and attain better STOL performance with the longer flaps.

Spencer already had proven the spoilerons to be practical on a unique hydro-glider called the Drag-N-Fly, which, towed behind a

Spencer Amphibian Air Car is homebuilt version of the Republic SeaBee, both designed by P. H. Spencer.

speedboat, isn't categorized as an airplane. The idea originated with a friend, Frank V. Nixon, president of Volpar, Inc., of Van Nuys, California, as a sort of fun flight trainer that you can tow aloft, then cut loose and glide back to water landing, using spoilers on the lower wings for roll control. Believe it or not, Spence first soloed Drag-N-Fly on his 80th birthday!

Chapter 12
Yesterday's Wings Today

The Experimental Aircraft Association recognizes two special categories of sport aircraft that involve either restoring or replicating those wonderful machines of yesterday's skies—Antiques (pre-1946) and Classics (1946-1950). A large number of EAA members do not concern themselves with modern-day homebuilt designs, preferring instead the nostalgia of flying as it used to be, when skies were less crowded.

In many an old barn, hangar, or backyard lie the ancient bones of ships that were queens of the sky in their time, and these are highly prized objects of a search by a growing number of enthusiasts seeking at least a few authentic pieces of some historic craft.

There's still another category EAA recognizes—the Warbirds that fought in the sky from 1937 onward, and the bulk of those are, obviously, World War II fighters and bombers, such as those that comprise the amazing Confederate Air Force at Harlingen, Texas.

In 1953, a group that called itself the Antique Aircraft Association was born, the same year that the EAA got started, and today there are some 5000 members, with chapters across the land. They categorize oldies a bit differently than the EAA does—an Antique to the AAA had to be built before 1935; from 1935 up to World War II was the Classic period; there's a special division for World War II War Birds, and from 1946 to 1950 is what they designate the Neoclassic period.

An avid enthusiast of the World War I period is Joe Mason, of Woodland Hills, California, who fell in love with the De Havilland DH-2 of World War I vintage, an unusual single-seater biplane with pusher propeller and machine gun mounted forward. Joe found a picture of a DH-2 in a magazine in 1965 and it immediately became his Garage Queen pinup. He dug up as many other pictures as he could find and set out to make his own construction drawings to build an 8/10th replica.

The project was a great escape from the cruel realities of today, with everything costing so much, and he devoted 8-1/2 years to the construction job. Finally, on March 1, 1974, after Joe figured he'd spent a mere $4,500, there she sat, ready to fly, complete with a 90-horsepower LeBlond radial engine. Empty weight was 762 pounds, gross weight 1222 pounds, wingspan 25 feet. Once in the air he found she'd fly at 70 mph with lots of great wind-in-wires noise, and her climb was a good 400 fpm. Incidentally, he was limited to the 8/10th scale because that was as big as he could make it inside the family garage.

I had the pleasure of photographing Joe Mason in his lovely DH-2 at Mojave Airport near Los Angeles, before he donated it to the San Diego Aerospace Museum in 1979, to help them back on their feet after a disastrous arson fire that leveled the previous establishment.

This is the kind of dedication you'll find among builders and pilots of replica warcraft, the majority of whom belong to an organization called the Replica Fighters Association, whose president is Frank G. Weatherly, of 2789 Mohawk Lane, Rochester, Michigan. RFA's primary purpose is to coordinate the efforts of individuals attempting to reproduce through size reduction, flyable military-type aircraft while maintaining recognized safe construction techniques, methods, and materials.

A typical RFA member is Dr. George W. Clapp, a dentist of Olean, New York, who chose for his replica project a 7/10th scale Chance Vought F4U Corsair. Though not a Corsair pilot, Dr. Clapp found much encouragement and help from other RFA members, and got so enthused that he built his own 2000-foot airstrip in his own backyard. He also got involved in a Nieuport 28 replica construction effort and visited a West Coast builder, Jim Appleby, who with his lovely wife Zona runs a unique business called Antique Aero at Flabob Airport in Riverside, California.

My introduction to the wonderful world of replica warplane flying was something special, so let me tell you about it: There I was at three thousand feet over the Western Front, pulling up through a loop to get onto the tail of Baron Manfred Freiherr von Richthofen's deadly crimson Fokker Dr I Dridecker.

Joe Mason built this lovely DH-2 replica from a magazine photograph.

I glance over my shoulder quickly. We are alone in the sky—I and the Red Baron! The wings of my Nieuport 28 C-1 are taut as I reach out and pat my plastic machine gun. In a moment, now, I will be diving down on the unsuspecting Baron, who is waving at the pretty girls on the airport below.

Baron von Richthofen waving at pretty girls, with certain death on his tail? That's what I said!

I have a split second now, to arm my gun and dispatch him, and break his record run of 80 kills with the murderous Flying Circus Squadron. But, at the top of the loop, my Warner Super Scarab coughs and quits. Oil spurts from the filler neck. My goggles, my windscreen are coated. I am blind!

Oh, my, I think quickly, what will Squadron Leader Jim Osborne say? It's his Nieuport, after all! Maybe I can save the day with a lucky shot, but as I squeeze the handle of the plastic machine gun, no bullets come out. Not even bubble gum. What the hell???

This, as I said, was my introduction to the wonderful world of World War I replica aircraft flying, high above Flabob Airport. Osborne, boss of the Stolp Starduster Corporation, had entrusted his Nieuport to me on the promise I'd bring it back in one piece. It was an unforgettable flight, even if the Red Baron existed only in my imagination, which really got turned on as I crouched in the small cockpit of the gaudy little ship with its bright red cowl and camouflaged wings and fuselage.

Replica flying is one of the hottest things in the homebuilt movement today—instead of doing your thing in a fancy new fiberglass ship propelled by a Volkswagen engine, why not re-live the great old pioneer days of military flying as it took place sixty years ago?

Outfit yourself in helmet and goggles and grow a pointy moustache. Learn to hum *Over There* and *Mademoiselle From Armentiers, Parlez Vous.* Read up on how to execute a vrille, a renversement, or a retourment, the way they taught the chasse pilots at Pau.

Next, swagger out to the flight line in your leather puttees and whipcord flying breeches, wrap a silk scarf around your neck, and swing a booted foot into the cockpit of your fighter plane, powered with a rotary, or at least a radial engine, and listen to the gals squeal.

That's the scene today around a growing number of airports where the nostalgia factor of replicas rates higher than the charisma of an executive turbo twin. The actors in this new drama of the sky are mainly ex-military jocks from World War II days, who've made their pile, and are looking for action reminiscent of their youthful days. This word comes from Jim and Zona Appleby, who built Jim

Jim and Zona Appleby built this Neiuport 28 replica.

Osborne's Nieuport 28, with Jim welding and Zona stitching on the fabric, a husband-wife team doing exactly what they wanted to do.

Back in their Antique Aero hangar rest the bare bones of an Albatros D5A, waiting to be powered with a six-cylinder, 180-horsepower, water-cooled Mercedes engine, and out front is parked a gloriously red Fokker Dr-1 Dridecker with a 145-horsepower Scarab, next to a recently-built Sopwith Pup replica.

The Applebys enjoy putting on exciting airshows with their World War I replica planes, complete with plastic machine guns that flash red neon lights in their muzzles and a sound track that goes BANG! BANG! BANG! when Jim pulls the trigger. It's enough to scare anybody to death.

Where the Applebys are strictly custom builders of World War I aircraft replicas, a number of other people are in business selling plans and kits of oldtime fighter planes of various sizes. Ken Thoms, of Santa Paula, California, a former advertising artist who loves airplanes, does business as War Aircraft Replicas, offering scaled-down plans of such aircraft as the German Focke Wulf 190, Hawker Tempest II, Republic Thunderbolt P-47D, Curtiss Hawk 75 A-3, Mitsubishi A6M5 Zeke 52, Macchi C.200 Saetta (Lightning), Lavochkin La.5FN, Grumman F6F Hellcat, Fokker D.XXI, and F4U Corsair.

Construction is composite, using plywood and polyurethane foam wing ribs covered with high-strength fabric and epoxy resin, and a fuselage built around a plywood wooden box, its contours made from carved polyurethane foam, fabric and epoxy covered. Retractable landing gear is available. Initial prototype was the F/W-190, which first flew in 1974.

In Athol, Idaho, Walter Redfern sells plans for two delightful fighters, the Fokker Dr-1 and the Nieuport 28, similar to the show planes the Applebys turn out. He uses a 145 horsepower Warner in the Nieuport, but as they are scarce, he says you can stick a 150 or 160 horsepower Lycoming under the round cowl and not spoil the looks. Redfern's plans have all formers, ribs, and fittings in full size, and he'll turn out special parts for you on order.

More than a decade ago a Roumanian-born Parisian named Marcel Jurca really got the whole World War II replica fighter-plane action rolling with a sleek machine he called, whimsically, *Gnatsum*—Mustang spelled backwards. It was 1967 and Jurca, who flew twin-engine Henschel 129-B2's for the other side in World War II, simply had to get something out of his system, as a frustrated ex-fighter pilot. Ever since 1949 he'd been building planes, beginning with the popular little Jodel.

Walter Redfern sells plans for his Fokker Dr-1 triplane.

Jurca's plans are far out—they come in a single 25 by 32 inch sheet followed up by sets of drawings for the various components, complete with humorous cartoons and balloons ("Don't scratch your head . . . BUILD IT . . . FLY IT . . . and you'll be a true pilot!").

On his fifth design, the MJ-5 Sirocco, he added a bump to the rudder (special feature for retract tailwheel possible) with the note: "This half-egg serves to—nothing. But gives the aircraft personal character! We call that in French "Pason-Nose" (Cul de Poule)." His first original design was the MJ-2 Tempete, which was highly successful as a sport plane in Europe, and then came MJ-5, the Sirocco, a low-wing two-placed sport plane. Gnatsum got the replica movement rolling as a scaled-down P-51D of either 2/3 or 3/4 scale proportions, and next came a 3/4 scale FW-190, an Me-109, the Spitfire, and a Curtiss P-40.

At presstime, two MJ 7 Gnatsums had flown—one built by Capt. W. T. "Butch" Foster and the other by Sturgeon Air Ltd. and called the SAL 2/3 P-51, both in Canada—and roughly a score more Jurca mini-fighters were being completed, including a third MJ-7 and an MJ-10 Spitfire in England; a 3/4 scale Messerschmitt MJ-109 in Reno, Nevada, and a 3/4 scale Curtiss P-40 Warhawk in Carson City, Nevada. Capt. Foster, a veteran fighter pilot with 9000 logged hours flying time, considers Jurca "a real pioneer in the replica field

Capt. Butch Foster flies his homebuilt Marcel Jurca Gnatsum.

who should receive just credit for initiating the WW2 replica fighter concept."

From England comes good news—John Isaacs, designer of a lovely 7/10 scale replica of the Hawker Fury biplane fighter of the 1930s, has flown a newer replica, the Isaacs Spitfire, a 6/10 replica of the famed RAF fighter, stressed to +9 and −4.5 G's for aerobatics. The Isaacs Fury first flew back in 1963 powered by a 65 hp Walter Mikron engine, and later was modified to take a 125 horsepower 0-290 Lycoming flat-four engine.

The Hawker Fury was born in 1929 as a single-seat biplane fighter of the Royal Air Force, and later was used as a trainer when bigger, faster warplanes came along. Powered with a 600-horsepower liquid-cooled inline Rolls-Royce Kestrel engine she could hit 220 mph. In Watervliet, Michigan, Robert E. Lohr, a carpenter by trade, saw a picture of the Hawker Fury in one of his son's school books and immediately sent to Professor Isaacs for plans for his version.

In 1975, the Lohr Isaacs Fury was almost completed in the family garage when an out-of-control snowmobile crashed into it and demolished the plane. It took five months to repair the damage, and finally, in 1978, she was flying. With a Lycoming 0-235C of 108 horsepower, his Isaacs Fury replica cruises at 93 knots IAS and is fully aerobatic.

The Isaacs Spitfire uses the Continental 0-200 engine of 100 horsepower driving a French Ratier metal prop, and can hit 150 mph TAS in level flight. Isaacs operates his prototype from a 700-foot grass strip, and in 1976 it appeared at the EAA Fly-In at Oshkosh after being transported over the ocean to Washington D.C., and then flown the 860 miles to Oshkosh.

Fred Sindlinger, of Puyallup, Washington, fell in love with the World War II Hawker Hurricane IIC fighter, and in 1969 began drawing plans for a 5/8 scale replica for homebuilders. Its first flight was in early 1972, and today plans and some component parts are available from him. A dozen or so are under construction in Africa, Australia, and Germany, and another is flying in Florida, owned by Ronald Nowling, an ex-USAF mechanic. Construction is of wood and fabric and the gear retracts. The canopy slides back for access to the single seat, behind a 150-horsepower Lycoming 0-320. The craft weighs 1005 pounds empty, can climb at 1850 fpm and hit 200 mph wide open.

Besides replica fighter planes, the Experimental Aircraft Association recognizes restored original military aircraft and maintains a special EAA Division, the Warbirds of America, Inc. A former Warbirds president is Rudy Frasca, head of Frasca Aviation, Champaign, Illinois, developers and manufacturers of a line of aircraft flight simulators.

Frasca has two great loves, besides his wife, Lucille, and his business—airplanes of all kinds, and his growing family of eight kids,

War Aircraft Replica's half-scale F/W-190 in action.

five boys and three girls. Joe, John, and Tom, all in their 20s, are pilots; Bob, a teenager is a student flyer, and David, 10, an aircraft modeler.

Among the Frasca fleet of flying machines are (or have been) a couple of Spitfires Rudy bought in India; one is on loan to the EAA Museum, and the other being restored in their factory building. There's a T-34 trainer that's getting a new Continental 285 horsepower engine installed, a Mooney Mark 21, Cessna 170B, a Cassutt racer, a Great Lakes bipe, and a Stinson Gullwing.

And then there are the big bores—Rudy's Curtiss P-40E and an FM-2 Wildcat that son Joe loves to fly in formation with his dad at airshows. Frasca started off with the Wildcat, which he bought from a couple of airline pilots in Atlanta, Georgia, in 1968, one of the first WW2 fighter restorations to fly.

In one airshow at Windsor, Ontario, Canada, the gear collapsed on a landing, and the repair job came to $16,000, one reason you don't see many real fighters around today! I can recall buying a Vultee BT-13 "Vibrator" basic trainer for only $600 right after the end of the war, with only two hours on the engine, but those days are gone.

Frasca has had expert help restoring his warbirds, fellows like Bill Ross, Harry Drummond, Ray Middleton, and Max Huffman, and now son Joe helps maintain the warbirds he flies as a student A&P, and buys his own avgas. Comparing the P-40 and the Grumman FM-2, Rudy says the most glaring difference between them is their operating cost, with the P-40E five times as expensive to operate as the Wildcat.

Both fighters are beautifully restored with authentic military paint schemes in polyurethane paint, and only in long range cruise mode does the operating cost of the P-40E favorably compare with the Wildcat. Flying at low-speed economy cruise, the Allison-powered P-40E, even with platinum plugs, cannot operate without the hazard of plug fouling, whereas the Wright 1820-56 can. In normal cruise the P-40E burns from 37 to 50 gph and the Wildcat only 30 to 37 gph.

All of which is reason enough to go to a scaled down fighter replica with an engine from 100 to 300 horses that won't eat all the hay in the barn on one flight!

Chapter 13
The Whirlybirds

According to a recent FAA survey, the fastest growing segment of civil aviation is rotary wing aircraft, soon expected to burst the bonds of complex technology and soar skyward by the thousands, primarily as a practical replacement for small corporate jets and turboprops. Their ability to take off and land vertically, hover, and operate from minuscule pads makes them ideally suited for short flights.

Some 7000 helicopters are now flying in the United States, and the majority of these are derivitives of military designs. But now, in expensive rotor blades are being made from composite materials that can last the lifetime of the aircraft, rather than being replaced after 500 to 600 hours of service.

Despite competition from German, Japanese, Italian and French manufacturers, as well as other firms in the United States, Sikorsky Aircraft is out in front with its 12-passenger S-76 Spirit helicopter that cruises at 166 mph and has a 460-mile range. Nearest competitor is Bell's 222, a 10-place aircraft.

Reflecting this unexpected boom in commercial rotorwing flying was an FAA program initiated in 1979 to overhaul its regulations concerning operation and certification of helicopters, which heretofore have been based on those for fixed-wing aircraft and do not reflect the unique capabilities of the whirlybirds, including vertical movement and solid IFR flight.

Among the smaller firms moving into the action are Robinson Helicopter Company of Torrance, California, with their R-22, and Eagle Helicopter Corporation, of Grand Island, Nebraska, with their air-jet rotor and tip mounted jet Eagle helicopters. By applying thrust to be rotor tips, Eagle's design eliminates the conventional helicopter's tail rotor used to offset the reverse rotational force produced by the rotor spin. Unique are the small jet engines at the end of each rotor blade of the Eagle helicopter, power units designed by Eugene M. Gluhareff, of Gardena, California. They burn liquid propane and the only residue is C02 and water, and there is no compressor or moving parts.

With such exciting developments in the world of commercial rotorcraft, it was only natural that similar exciting advancements would appear in the homebuilt aircraft area. Outstanding for its simplicity of construction and operation is Martin Hollmann's HA-2M Sportster gyroplane; but before we go further, let's go back to 1923 when a Spanish inventor, Juan de la Cierva y Cordonia, made history's first successful flight in a thing called an autogiro.

Rather than whirling the lifting arms with power applied to the central shaft, Cierva's rotor blades were hinged at the center, permitting them to change pitch or "flap" as they spun around, the spinning produced by air pressure as the craft was pulled forward by a conventional engine and propeller.

The rotary wing concept goes back to the time of Leonardo da Vinci, who called his proposed machine a *helix* (Greek for spiral), but not until the 1920s did the concept prove practical. In 1925 Cierva took his machine to England, where Captain Frank T. Courtney test flew it for the British Air Ministry. Soon the brilliant Spanish homebuilder had signed a deal with A. V. Roe & Company to form the Cierva Autogiro Company of England. In 1928, he amazed the world by flying his machine across the English Channel.

Early autogiros were in effect conventional airplanes with four rotor blades mounted on a pylon, but in 1934 an American firm backed by a wealthy Philadelphian, Harold F. Pitcairn, the Autogiro Company of America, introduced a wingless autogiro whose stability was controlled by variable-pitch rotor blades. Amelia Earhart flew one coast to coast. The autogiro was inefficient and gulped fuel, and soon was replaced by the helicopter. In 1939 Igor Sikorsky made the first successful flight of a helicopter in the western hemisphere in a contraption he called his Ugly Duckling.

Since that time, with the stimulus of military demands for helicopters during World War II, that kind of rotorcraft has dominated the commercial rotorcraft scene, and found its place on urban star-route passenger and mail short-haul runs.

One day in 1939, two men, whose families were neighbors in Kiev, Russia, met face to face—Igor Sikorsky and Dr. Igor B.

Hollmann Sportster is only two-place homebuilt gyrocopter.

Bensen, P. E. —and for the latter, the meeting changed the course of his life. Bensen hoped to join Sikorsky in helicopter work, but World War II intervened. Instead he joined General Electric Company as a research engineer in the field of rotary-wing development.

In 1951 Bensen left G.E. to become chief research engineer for the Kaman helicopter firm and in 1953 finally formed his own business, the Bensen Aircraft Corporation in Raleigh, North Carolina. His goal: develop a mass-producable "People's Flying Machine" capable of door-to-door transportation. However, anticipated heavy military orders for a ramjet-powered Model 5 helicopter failed to materialize when the Korean War ended.

Coincidentally, Boys Life magazine published a photo of Bensen flying his first Gyro-glider, described as "so simple to build and fly, any high school boy could master it." Overnight Bensen found himself swamped with requests for plans and kits for a homebuilt Gyro-glider. So, he removed the ramjets from his B-5 helicopter and converted it to a Gyro-glider. Bensen estimates that some 4000 Model B8M Gyrocopters were built since then, with half that number still flying, making it the most popular of all homebuilt aircraft.

The Model B-8 Gyro-glider, which started the boom for Bensen, is an unpowered rotor kite you tow behind a car, then cut loose for free flight. Several hundred kits and plans have been sold for the craft, which required no pilot license to fly, being classed as a kite rather than an aircraft. A free-turning two-bladed rotor is universally mounted atop the craft's frame, and is normally tilted by an overhead stick to control direction of flight. Rudder pedal controls move a tail rudder.

The Bensen Model B-8M Gyrocopter, first flown in 1955, is a powered autogyro version of the Bensen Gyro-glider designed for home construction. Fitted with floats it becomes the Hydro-Copter. Bensen used the name Gyrocopter to reflect the dual nature of the basic Bensen design, combining flight characteristics of the GYRO-plane and heliCOPTER.

Its rotor blades are set in autorotative pitch, yet the rotor is powered in flight. In this configuration it cannot hover but can make vertical takeoffs and landings. The craft, says Bensen, has no collective control stick to worry about and is already capable of a power-off glide in the event of engine failure.

Gyrocopter landings are executed power-off or on partial power, and cyclic flare alone arrests forward motion when the rotor is tilted aft to its full travel, but the recommended technique is to land with a slight forward roll to eliminate the chance of side drift during touchdown. The Bensen Gyrocopters are actually roadable and can be driven down streets like a motor scooter to achieve "door-to-door" transportation.

The Model B-8M Gyrocopter can use the 90-horsepower McCulloch 4318E flat-four two-stroke engine or the 64-horsepower Volkswagen 1600cc powerplant, and can be fitted with an optional mechanical rotor drive. Other options include larger diameter rotor, an offset gimbal rotor head, a floor-type control column (instead of the overhead stick), dual ignition, nosewheel arrester, and pontoons. A derivative, the Bensen Super Bug, uses a twin-engine installation to spin up the rotor prior to takeoff.

In 1976, Bensen succeeded in developing a truly hovering machine, the Model B-18 Hover-gyro, described as a gyrocopter with two coaxial counter-rotating rotors. The upper rotor is larger in diameter, and its blades are set in autorotative pitch. The lower rotor has automatic collective pitch control which responds to engine power. Both rotors are power driven, and a smaller pusher engine/propeller provides forward thrust and yaw control.

Incidentally, a curious precursor of the Hover-gyro was patented in 1910 by Gustave H. Brekke of Seattle, Washington. The Brekke machine had two counter-rotating four-bladed rotors made of bamboo and silk, the lower rotor wider than the upper rotor. The

Dr. Igor Bensen poses with his prototype gyrocopter Kitty Hawk.

blades of the lower rotor were pitched opposite to the upper blades and driven in an opposite direction to cancel out all torque. There is no record that it ever flew.

For all the fun of Gyrocopter flying, NTSB accident reports reveal that they are perhaps the most dangerous of all homebuilt aircraft, largely due to improper operation, such as a push-over at the top of a steep zoom, creating a negative G load and loss of the rotor. Other accidents have resulted from high-speed taxiing in strong wind conditions, producing a roll-over; a delayed recovery from a rapid descent making a flareout landing impossible; and tip-overs in flight.

Several other rotorcraft have appeared on the market, some successful, others not. One successful machine is the JT-5 Autogyro designed by Jukka Tervamaki of Helsinki, Finland, a single-seater with a 23-foot rotor span and powered with a 75-horsepower converted 1700cc VW engine by Peter Limbach. The two-blade semirigid rotor is of fiberglass and epoxy resin construction over polyurethane core, the blades of constant chord and NACA 8-H-12 airfoil section.

The rotor pre-rotation system consists of a V-belt clutch, 90-degree gear box, sliding universal shaft and an inertia-operated

Bendix drive, with the pre-rotation control lever pivoted to the throttle lever for simultaneous use by the pilot's left hand.

Fin and rudder are of fiberglass reinforced epoxy resin with rigid PVC-foam ribs and Courtaulds carbon-fiber stiffeners. Horizontal tail and tip fins are of fiberglass sandwich construction with a honeycomb core. The JT-5 thus is a pioneer in state-of-the-art composite materials construction, which Tervamaki adapted from his experience as project manager of an all-fiberglass glider tug plane, the PIK-19, for Helsinki Technical University. JT-5 also is a development of an earlier model, his ATE-3. It can climb at 540 fpm and cruises at 81 mph.

For a moment, let's look at homebuilt helicopters and see how they compare with gyrocopters. Gyrocopters obtain lift through the action of air flowing upward through the rotor blades, producing autorotation. Hence, should the propulsion engine quit in flight, a descent under gravity power continues autorotation and permits the gyrocopter to effect a safe landing. By contrast, the helicopter's blades are mechanically driven, and in the event of engine failure the "climb pitch" angle of the rotors, roughly 11 degrees, would quickly stall them, unless the pilot swiftly reduced the pitch angle of the blades to one that would provide safe autorotation. This transition may cause a disastrous loss of altitude unless the pilot reacts quickly. One newer homebuilt helicopter to appear is the Scheutzow Hawk. Its designer, Webb Scheutzow, of Berea, Ohio, claims it has outstanding stability and maneuverability, plus a fast cruise speed. The Hawk followed the appearance of the Scheutzow Bee, a new type helicopter with a rotor head in which the blades are carried on rubber bushings. The Bee is a light, side-by-side two-seater using a 180-horsepower Lycoming engine. Scheutzow's patented Flexhub incorporates elastomeric bearings with an offset flapping hinge to reduce vibration and provide better control power and precise response in all maneuvers. The Bee has a 27-foot rotor diameter and cruises at 80 mph.

Back in 1966, a 28-year-old helicopter design engineer named B. J. Schramm introduced to the homebuilt market a single-place helicopter called the Javelin, hoping to sell it either fly-away or as a prefab product for under $10,000. Response was such that in 1967 he set up a company called RotorWay Inc. of Tempe, Arizona, to market a production version called the Scorpion, with plans and kits for amateur constructors.

Subsequently a new model replaced the single-seat model—Scorpion Too, a side-by-side two-place model, that is powered with

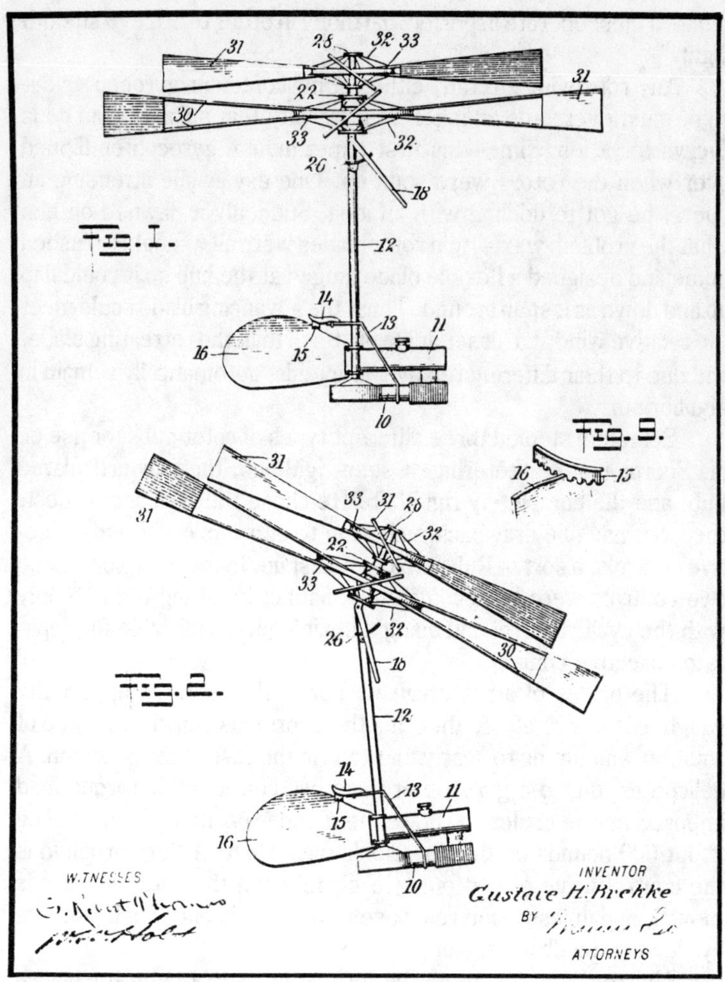

1910 patent of G. H. Brekke's co-axial rotorcraft.

a specially developed RotorWay horizontally-opposed engine, the RW-100, intended for the homebuilder market.

RotorWay provides comprehensive plans, technical assistance, preflight training, and kits, either complete or modular. Plans and rotor blades also are available. Says Schramm: "A helicopter is essentially a series of vibrations all of which are made to function in harmony with one another." (Sounds a little like a bellydancer with the hives?)

"Three major flywheel effects are present," he goes on. "The rotor-blade system, the tail-rotor system, and the powerplant system. All must operate in perfect harmony in order to achieve smooth flight."

Any rotorwing aircraft, either of helicopter or gyrocopter design, must overcome an inherent instability that puzzled Juan de la Cierva for a long time—his first experimental gyrocopter flipped over when the rotors were spun up. One day, while attending an opera, he got to fiddling with an idea. Suddenly it dawned on him what the problem was—rigid rotor blades were all wrong! He rushed home and designed a flexible blade hinged at the hub so it could flap up and down as it spun around. Thus, the advancing blade could meet the relative wind at a lesser angle of attack than the retreating blade, and due to their differing relative airspeeds, automatically remain in equilibrium.

Schramm studied three different types of rotor hubs for use on his Scorpion—the teetering or semi-rigid hub, the fully articulated hub, and the completely rigid hub. He chose the first one as best, though it had one drawback, in having to intermix cyclic and collective controls, a sort of Rube Goldberg lashup. In the Scorpion, these two controls were kept completely distinct by tilting the hub only with the cyclic control and using a flexible push-pull cable for separate collective control.

The blades of a rotorcraft do not in themselves support the weight of the aircraft. Rather, it's the enormous centrifugal force of rotation, amounting to nearly four tons in the case of the Scorpion. A helicopter, unlike a gyrocopter, also has considerable torque load imposed on the blades to rotate them under positive pitch, totaling about 600 pounds on the Scorpion's main shaft. A third problem is the control force placed on each blade when the cyclic control is moved, and unless blade response is identical in each blade, severe cyclic stick feedback results.

Unlike the gyrocopter, the helicopter uses a tail rotor placed outside the main rotor disc area to maintain directional stability, and in Schramm's design, the tail is reminiscent of the stinger of a desert scorpion for which it was named.

It all adds up to a unique way to fly in a homebuilt. Says Schramm: "At best, the freedom of flight in a helicopter is difficult to describe and can only be truly appreciated by experiencing this exhilerating sensation yourself. To build a personal helicopter with your own hands and then experience the excitement of flight in the craft is a rewarding experience which few other projects can compete with."

We now come to what many people feel is the ultimate homebuilt rotorcraft in safety, inexpensiveness, ease of construction and piloting—Martin Hoilmann's two-place HA-2M Sportster gyroplane, which can be built for around $6,000, or from one-half to one-quarter the cost of building a Scorpion Too.

Born in Germany in 1940, Martin is the son of a noted German scientist who escaped East Germany and came to America under Operation Paperclip, a government project to provide refuge for talented scientists from the Communist take-over after World War II. An avid surfer and backpacker, Hollmann, who stands 6 feet 7 inches, was turned onto gyroplanes in 1968 when his instructor at San Jose State University introduced him to Cierva's C.30 gyrocraft.

Hollmann won a Student Fellowship Grant and built and tested successfully a 20-foot diameter fiberglass laminate rotor, followed in 1974 by the design, construction and flight testing of a 1050-pound gross weight two-place gyroplane. Hollmann completed his schooling that year and won his Master's Degree in Mechanical Engineering at Florida Tech, then moved to California, where he works for Lockheed Missiles & Space Company at Sunnyvale as a senior design engineer. Oh, yes—he brought his gyroplane with him, towed behind the family Volkswagen beetle at 55 mph!

Today Hollmann devotes all his spare time to making parts for homebuilders and preparing newsletters to help them build and fly the only two-place gyroplane for amateur constructors.

Curiously, Hollmann's dedicated research program helped others get started in the gyroplane business well ahead of him, when he developed a special rotorblade system capable of providing a low disc loading of 1.8 lb/sq. ft. He started by using a fabricated 2024-T8511 aluminum leading edge extrusion and 2024-T3 alclad skin riveted structure rather than a laminated fiberglass structure. To keep the weight down, a low solidity ratio of 0.035, common to gyroplane rotors, and a teetering, two-bladed rotor system were chosen. The NACA 8-H-12 airfoil was used and the blades designed to meet FAR Part 27 requirements.

At this point, Hollmann heard that a firm in San Diego was having problems flying their gyroplane, called the Boomerang, on smaller Rotordyne blades, and built them a set of his HA-27 rotor blades. They worked fine. Meantime, in England, Campbell Aircraft Ltd. ordered a set of the HA-27 blades for their 1000-pound gross gyroplane, the Cougar, which subsequently was flown to the 1972 Paris Air Show.

Finally his HA-2 Sportster was built in Florida, towed to California, and modified with a bigger engine. Says Hollmann: "Generally speaking, twice as much horsepower is required for a rotorcraft, helicopter, or gyroplane, to lift a given weight, than for a fixed-wing aircraft." This dictated his purchasing a 150-horsepower Lycoming O-320 GPU engine to replace the 130-horsepower Franklin Sport 4B used in the HA-2 model. The new design was called the HA-2M.

In addition to installing a bigger engine, Hollmann decided that its performance could be improved by modifying the cockpit enclosure with better streamlining to decrease drag and improve the airflow into the propeller mounted behind it. Other improvements included lengthening the hub and rotor diameter from 27 to 28 feet and increasing the blade pitch by one-half degree. These two changes improved the ROC from 350 to 500 fpm.

Says Hollmann: "The Sportster's large, free-wheeling rotor allows it to be landed power off almost anyplace, and its high degree of maneuverability in the horizontal plane can only be matched by the helicopter. However, unlike the helicopter, it has a direct-drive propeller which pushes the aircraft through the air. It has no transmissions or drive trains, belts, or pulleys and is mechanically one of the simplest and safest of any aircraft. Unlike the helicopter, the gyroplane possesses both static and dynamic stability in pitch and yaw and its large rotor has a low control sensitivity. It has a wide speed range, normally from 20 to 90 mph, making it the ideal sport aircraft."

In an emergency power-off autorotation landing, the Sportster's approach slope is steep, about 45 degrees, with a sink rate of about 1000 fpm, permitting landings in small clearings surrounded by high trees. By proper flaring, achieved simply by pulling the control stick back, a zero ground roll can be achieved. There is positive control response at all flight speeds and since the rotor is always auto-rotating, the craft cannot stall or spin.

The Sportster utilizes a rotor prespin drive which allows the rotor to spin up to about 240 rpm. At this rotor speed the craft is accelerated from a stop to 45 mph for a no-wind takeoff in as little as 350 feet. The Sportster's rotor head control system uses a simple push-pull linkage permitting all aircraft control except for engine throttling to be achieved with the single control. Rudders are employed only to increase the turn rate—to make a turn, you simply push the control stick in the direction you want to go.

I had the pleasure of flying the Sportster with Hollmann at Fremont Airport near San Francisco Bay, and found it simple to maneuver and delightful to move about all three dimensions smoothly and easily. In fact, I felt it so safe and simple to operate I let my daughter, Toni, fly with Hollmann to experience a new thrill in sport flying!

Chapter 14
The Hover Lovers

In the homebuilt aircraft movement, something new is always turning up. If it isn't a jet-powered hang glider it might be a homebuilt rocketship, or a one-man helicopter you can strap on your back in Buck Rogers fashion. All these devices have actually been built, but there's another way to go that is fast becoming a worldwide fad—do-it-yourself Hovercraft!

These are small, lightweight surface skimmers that scoot along in ground effect, over water or land, and take you into areas of the world otherwise inaccessible to conventional land, air, or watercraft.

Largest supplier of homebuilt hovercraft components is Barry H. Palmer, an EAA member of Seattle, Washington whose machine is exactly what its name implies—FAN-TASTIC! Says Barry: "FanTastic barely ripples the water as it skims quietly along at more than 20 miles an hour, on an incredibly low six horsepower! Powered with an ordinary, mild-mannered garden tractor engine for outstanding reliability, and which anyone can build with simple tools for around $450 it will carry an adult and a youngster of up to 240 pounds combined weight over field, marsh, beach, or snow; up rapid rivers, and into hard-to-reach places where the great outdoors is at its greatest! For the fun of flying on a seven-inch thick cushion of air, it's FAN-TASTIC!"

His enthusiasm may be pardoned, for hovercraft actually are opening the doors to a whole new world of sport and exploration. The single-place Fan-Tastic first flew in 1974, with plans and kits coming out in 1975. Since then, close to 100 Fan-Tastic machines have been built, the amateur builders ranging in age from 15 to 65, and coming from all walks of life.

Fan-Tastics are distributed in the United States, Canada, and Australia, and in many other corners of the world. Says Palmer: "Although the prototype was constructed using an 8 horsepower Briggs & Stratton engine, and is still being operated with that original engine, builders are now installing the newer 11 horsepower Briggs & Stratton engines. Fastest completion of a Fan-Tastic was accomplished in one month of spare time, though from four to seven months is the average building time."

Five different engines have been tested on the craft, and as a result only twin-cylinder de-tuned snowmobile engines are now approved for installation. A 16-horsepower Briggs & Stratton engine is now installed on Palmer's machine. Currently, he's running tests with a still bigger machine that uses a Datsun B210 engine of 1.4 liters, and can be configured to sleep four people, with a galley and portable head, or just haul six people.

Fan-Tastic, also known as ACV-1-8, is an air-cushion vehicle, designed for high reliability through conservative drive and rotor systems and use of a mildly tuned powerplant. Noise level depends on engine muffling as the rotors are essentially quiet. Palmer selected an open or unconfined propeller for propulsion, because of the difficulty for the amateur builder in building a large diameter duct and in maintaining the tight blade tip clearances of ducted-fan type SEVs (Surface Effect Vehicles).

SEVs actually date back to the turn of the century in patent office files, but a practical technology has only evolved in the past two decades. In the 1960s in England, five firms turned to development of ground effect machines whose operation was based on use of an air curtain system invented by C. S. Cockrell. In this system a curtain of air is directed downward around the periphery of the vehicle, sufficiently to hold a low-pressure air cushion in place underneath, less than one psi above outside atmospheric pressure. One firm, Folland Aircraft, went to the principle of an induction pump with high-velocity air injected through narrow slots to entrain the larger volume required to produce the curtain airflow. Their research vehicle was called GERM (Ground Effect Research Machine) and was directed toward development of a general purpose Hovertruck with a five-ton payload.

Shortly after World War II, the U.S. Navy began a Mission Analysis Study of SES (Surface Effect Ships) craft, and concluded they held great promise within the 2000-to-3000 ton size for any missions assigned to surface forces, as well as special missions such

Palmer Fan-Tastic SEV (Surface Effect Vehicle) is the way to go!

as search and rescue. By 1972, a 100-ton SES craft had become operational. SES machines, however, are not amphibious like the SEVs; they ride on captured air bubbles trapped by rigid, partially submerged sidewalls and flexible seals fore and aft.

While big business and government agencies were pursuing this exciting new research area, a number of private individuals and small firms moved in to popularize SEVs as a thrilling new sport. In the 1960s, the Australian Air Cushion Vehicles Development firm initiated an R&D program for SEVs, and has since moved to Terre Haute, Indiana, as Neoteric USA Inc.

Like Palmer Aerosystems, Neoteric and other young companies found themselves riding a real boom, and a SEV fraternity, the Hoverclub of America, Inc., was organized, doing business at Box 234, Uniontown, Ohio 44684 (it costs only $10 to join).

Recently, Neoteric USA developed a second-generation hovercraft, Neova II, as a homebuilt you can assemble from a $3000 kit of four modules, base, ducts and controls, machinery, and skirts (engine and outer body are optional). A good scrounger can build one for under $1500. Neova II has a unique thrust reverse system than enables it to back up. The craft is 14 feet long by 7 feet wide and travels at 35 mph carrying two persons of average weight. It has a simple chain-drive transmission and can handle any engine of around 46 horsepower (Neoteric uses the 1600cc VW engine).

Engineered with safety in mind, Neova II has skids underneath and the hull contains 150% bouyancy, with its curved fiberglass body resistant to accumulations of spray and debris. An integral siphon system eliminates any interior water buildup, and transmission, fans and engine are enclosed for safety.

Neova II flies over land and water on a 1/8 inch air cushion at 35 mph, floats like a boat in water at rest, and is highly maneuverable. J. Christopher Fitzgerald, Neoteric USA president, says the craft appeals to hunters, fishermen, farmers, construction engineers, and folks who just like to explore inaccessible areas. Recently, Neova II has been fitted with a fiberglass construction kit that lets the homebuilder assemble it in only 200 hours, instead of 1400 hours using plywood construction. A higher-performance three-seater, the Neova Super Sport, is available but is sophisticated and expensive—around $25,000.

By contrast, Barry Palmer's Fan-Tastic can be built for as low as $350 as an excellent back-country exploration machine. Says Palmer: "If a week in the wilderness is what you want, 20 to 30 mph cruising speeds of Fan-Tastic Two will get lots of gear into your base camp and still provide wetlands transport for two."

Fan-Tastic Two is unique, in that the air cushion is divided into a low-pressure forward section and a high-pressure aft section to get away from bow spray and leave a smoking mist high in the air behind. Fan-Tastic Two costs from $600 to $1200 to build, including engine, depending on the availability of used parts. A special section is provided in this craft for those willing to sacrifice a little reliability and install a snowmobile engine for better performance. If derated from 35 to 21 horsepower, says Palmer, they should provide adequate service.

Palmer has driven the prototype for thousands of miles over the rivers, lakes, swamps, tidal flats, and the waters of Puget Sound, some trips extendi ıg more than 200 miles. With the 16-horsepower four-cycle industrial engine, Fan-Tastic Two weighs 825 pounds gross, is 13-1/2 feet long and about 8 feet wide. Top speed over water is 32 mph, and over ice 38 mph.

There's no big deal in carving your own rotors for Fan-Tastic, says Palmer. The lift fan can be made from two lengths of 1-1/2-inch by 3-inch by 36-inch balsa blocks, along with some scraps to form the thick hub area. Detailed instructions for carving the blades are given in the ACV-1-8 manual, for both the lift fan and the propeller. After carving, the blades are glassed and resined, creating an outer shell that takes the structural loads.

Neova II Hovercraft opens a whole new world of fun.

The fun comes, of course, after the craft is all assembled and the engine tuned up. The SEV can be transported atop your car to the launch site and set up in five minutes. One morning at 6 a.m., Palmer launched his Fan-Tastic Two from his trailer ramp into Lake Washington, then headed northwest at 25 mph over glassy, lumpy seas. He negotiated a series of locks and then headed through open water for Whidbey Island, raising flocks of waterfowl.

Crossing eight miles of open water, he headed across a tidewater stretch, the Skagit Flats, and a vast coastal marsh wetland covered with tall grass. Then he went up the Stillaquamish River for an overnight campout, cruising at just above hump speed, often leaving the river to avoid swimmers by flying over gravel beds and sandbars. The trip covered 190 miles and ended with a broken fan belt.

"There's a lot more to surface skimming than racing on closed courses in sight of rescue at all times," says Palmer. "There is a whole, wide world of wetlands in most areas, sometimes surprisingly close to civilization. It's possible to cross the entire North American continent north to south via SEV without portaging, and east to west with very little. There's so much to be explored—that's what Fan-Tastic SEVs are all about!"

One of the larger SEV's is Universal Hovercraft's UH-18—it's 18 feet long, 8 feet wide, and weighs half a ton empty. Payload is 1200 pounds, hover height 12 inches, height of cushion 6 feet, maximum gradient 30%. Universal's Bob Windt, who estimates there are some 3000 SEVs in the United States, hopes to traverse the entire length of the Mississippi River some day by SEV.

Windt initially designed UH-18 as a twin-engined craft but switched to a single Corvair engine and Banks propeller, with a rear-lift fan and belt-drive to the prop. This reduction unit solved a high-rpm scream from the Corvair when used in direct drive. Later, he moved the lift fan back to the front again, and UH-18 can now fly over the waves at a mile a minute before bag tuck and danger of a plough-in.

Another SEV designer, Raymond Kemp, of British Sports Hovercraft in Tustin, California, in 1978 demonstrated his SEVs Starduster and Sea Mouse to the Navy in San Diego Harbor. He recalls:

"The demonstration flight was an absolute dream. In brilliant sun the two craft came around the point streaking in front of the grandstand looking, as a Naval Commander said, 'like ice cream on velvet.' As they went into spin and turn maneuvers, with the two craft interweaving and crossing each other, the TV people descended on us with requests for more footage."

SEV buffs resort to self-policing, as in the overall homebuilt aircraft movement, to insure safety and compliance with environmental standards. Says one SEV pilot, Dennis Benson: "Noise levels are very critical to hovercraft operation. They may not be safety related, but limits are needed so that we can continue to operate here in the Lower Forty Eight where people are beginning to outlaw noisy things. Even the boat and airboat people engaged in outright racing are finding that they must lower their noise levels. We are not associated with the Sierra Club, but we feel that the hovercraft, with its ability to enter and leave terrain leaving no mark of its travels, would fit very well into their scheme of things."

Chapter 15
The Featherweights

High on a windy hill near the bedroom community of Sylmar, California, a line of brightly colored butterflies awaits their turn to launch into space. They are mostly Rogallo Wing hang gliders, though some are more sophisticated Icarus biplanes and fragile-looking Quicksilvers. One by one their pilots check the freshening breeze, then sprint a few steps and lift off into glorious, free flight, to ride the slope winds with chicken hawks and gulls.

Over the San Gabriels, at a place called El Mirage Dry Lake, on California's Mojave Desert, other similar butterflies await their turn to launch, but there is no hill—only the parched, cracked alkali mud of an ancient lake, flat as a billiard table. And once more, one by one, the pilots launch themselves with a few running steps, but with a difference—these craft have tiny engines pushing or pulling them along to reach liftoff speed of maybe 21 mph.

The latter aircraft are an entirely new breed of sport planes, variously called, simply, powered hang gliders, or maybe ultralights, or minimum airplanes, or featherweights. Literally thousands of these little machines are flying today in all parts of the world in a virtual explosion of new-found freedom, albeit reminiscent of the early days of man-flight, when the Wrights and Glenn Curtiss and Louis Bleriot and other early birdmen first struggled aloft in powered planes.

The reasons for the explosiveness of the movement are simple—it enables young people to get airborne for a few hundred dollars, instead of paying maybe $25,000 for the dubious privilege of owning a stubby little Cessna 150 trainer. Again, it enables hang glider pilots to fly from virtually any launch site, without depending on slope winds blowing up a mountainside to get airborne safely.

The concept of ultralight flight goes back to Leonardo da Vinci in the 16th Century, when he experimented with flexible wings and once described how birds soar along ridges supported by what he called *reflex* winds. Today's sky surfers call them slope winds, of course. They are particularly brisk along California's coastal range, blowing in off the ocean at an average velocity of 10 knots. Sweeping through the tall grasses of the green hills, these slope winds can support powerless hang gliders for hours on end.

Hang gliders come in two basic varieties—the flexible wing type that intrigued Leonardo, and the rigid wing variety pioneered at the turn of the century by an American bridge engineer named Octave Chanute, whose design was "borrowed" by Orville and Wilbur, who hung on an engine to achieve powered man flight. That's just what the kids are doing today—hanging engines on hang gliders!

Three quarters of a century after Chanute's time the revival of hang gliding got under way in Southern California when a high school teacher named Jack Lambie helped his sixth-grade class build a Chanute-type biplane glider with a 28-foot span. It cost them only $24.95 and was made from scraps of wood, plastic drop cloth, and piano wire. They named her Hang Loose and were delighted to watch their teacher fly it down a coastal slope near San Juan Capistrano Mission.

Other biplane hang gliders quickly followed. One was designed by a 16-year-old youth named Taras Kiceniuk, who named it Icarus, after the legendary Greek youth who flew too close to the sun and melted his wax wings. Next came the flex-wing paragliders that combined the billowing canopy of a parachute with rigid V-shaped spars, developed in 1958 by a NASA/Langley Research Center engineer named Francis M. Rogallo. The Army and Air Force tested the Rogallo gliders for cargo drops and for recovering jet aircraft pilot ejection capsules.

There was some experimental work done on powering these Rogallo gliders, but they were first adapted for sport flying down in Australia where one devotee, Bill Bennett, used them to get airborne while water-skiing behind speedboats. Intrigued by the Australian's success, two San Francisco Bay area water skiers, Dave Kilbourne and Donnita Holland, both of San Jose, introduced the sport to America. Soon they abandoned the boat-tow launch method and took to the hills. Donnita became the nation's leading hang glider aviatrix, with several thousand flights to her credit, and Dave proved

John Moody's Easy Riser powered Icarus hang glider was first of booming power hang glider fleet.

how fantastic was the strength of the slope winds along Mission Peak, sailing aloft for more than an hour at a time.

As the sport caught on, amazing endurance records were set, and along the mountainsides of Oahu pilots have stayed aloft from sunup to sundown in powerless hang gliders. But what of the poor flatlanders, in America's Midwest, where there are no high mountains to soar over?

A young fellow named John Moody, of Milwaukee, Wisconsin, found the answer—he simply hung a McCulloch 101 chainsaw engine on one of Taras Kiceniuk's Icarus II hang gliders and presto—a "minimum airplane" was born! Moody's introduction of the new sport to the public was a near-disaster. At the 1976 EAA Oshkosh Fly-In, I watched John spiraling slowly skyward above the oak trees behind the camp grounds, the engine snarling like an angry hornet.

As I stared, the machine suddenly pitched up and over into an outside loop, then did two more gyrations somewhat resembling a Lomcevak, and—amazingly enough—John came to rest on his feet in a perfect two-point landing! Moody admitted he had banked too steeply in a 180-degree turn and rolled inverted, and was unable to shift his weight forward far enough to avoid a second tumble. He recalls:

"Nearing the ground, the craft tumbled a third time, at which time I shut off the engine by releasing my mouth-held shut-off safety switch, and made a normal landing. EAA officials three days later permitted me to fly again to show that the fault was with the pilot, not the design. I'd simply goofed and admitted it."

People were a bit skeptical of powered hang gliders after that, but Moody quickly recovered with an amazing altitude flight to 8700 feet MSL after a flatland takeoff from 690 feet MSL. Using an extra fuel tank, the flight was completed solely under engine power, with no assist from thermal, ridge, or wave action.

According to pioneer power-glider Moody, the most important aspect of the new sport is that it can make the sport far safer than in the old powerless days, allowing the pilot to gain proficiency at low altitude on calm days away from mountains and cliffs and associated turbulence.

"Another goal," he says, "was to make low-cost, completely portable flying available to the average homebuilder, and lastly I wanted to achieve these goals using an aircraft one man can handle, assemble, and transport, and teach himself to fly with reasonable safety. Among ultralights, with rigid wings, our configuration is unique with its parachuting type of stall, important to low time pilots flying foot-launching craft."

FAA officials initially stayed away from trying to regulate this strange new breed of aircraft, until one careless pilot took off from Playa Del Rey Beach into the takeoff corridor of jet aircraft departing Los Angeles International Airport (LAX). Radar monitors, startled, warned all traffic about the UFO, and were quite relieved when it finally disappeared from their scopes.

Officials wrestled with the problem with dichotomous indecision—they did not want to suppress the movement, yet knew the safety of all who fly is dependent upon some kind of air traffic control system, both to avoid the chance of a midair collision, or injury to the pilot and anybody on the ground should the craft tumble and crash.

At this point, the United States Hang Gliding Association was formed as a sanctioned representative of the Federation Aeronautique Internationale, which promulgated a Sporting Code for Hang Gliding and stressed the need for self-policing to avoid government crackdown on the new-found freedom of flight. According to FAI, the definition of a hang glider is "a heavier-than-air, fixed wing glider capable of being carried, foot-launched and landed solely by the energy and use of the pilot's legs.".

The FAA tended to accept this definition and carried it over to powered hang gliders—if it had to be foot launched and landed, it wasn't an airplane, but if wheels were required for landings and takeoffs, it was. Some PHG (Powered Hang Glider) builders had attached wheels to the pilot cage to make their craft easier to

Volmer Jensen hung an engine on his Sun Fun to make it the VJ-23E powered hang glider.

transport on the ground, but if they were not required for takeoff or alighting, it didn't matter.

Other designers simply accepted the definition and went along with it, providing a tricycle gear for getting off and on the ground, but also providing room for the pilot's feet to hang down for "two-point" landings. Ken Striplin went this route with his FLAC (Foot Launched Air Cycle) rigid wing monoplane PHG, installing "bomb bay doors" under the pilot seat to open the fiberglass cockpit area for foot launching and landing.

John Chotia, on the other hand, simply shrugged and accepted the inevitable with his fine little Weedhopper PHG, a tractor-engined craft in which the pilot lies supine, feet forward above a trike gear, depending entirely on engine power to get airborne. Weedhopper thus could qualify as the "world's smallest aircraft" and is licensed to fly in the Experimental Aircraft category by FAA.

Like other homebuilt aircraft, you have to build more than 50% of the machine for it to qualify as an Amateur-Built Aircraft, and the pilot must possess at least a student pilot permit to fly it. As Weedhopper, like other such PHG's, is a single-place machine, the instructor who endorses your student permit for solo flight stands on the ground and watches, biting his nails and hoping—.

Weedhopper is the result of a number of design changes by Chotia, who settled on a single-surface wing with a tail group behind, and when he realized that all surplus McCulloch and other small

Mitchell Wing is one of the finest of powered hang gliders flying today.

engines were being snapped up, designed his own powerplants, the Chotia 460, world's first engine designed specifically for use in ultralights. The engine delivers 18-1/2 horsepower at 3500 rpm permitting a direct drive propeller installation.

Weedhopper's inherent stability comes from its wing dihedral and low CG. It can carry a 220-pound pilot at 50 mph or cruise at an easy 30 mph. The Chotia 460 engine is lighter for its displacement than other two-cycle engines, and its lower rpm means longer life, with a TBO of from 800 to 1000 hours. (Both the Weedhopper and the Chotia 460 engine are available for the homebuilder—see Appendix).

Curiously, the world's first PHG appeared back in 1974, when an inventor in Gardena, California, designed a small jet engine with no moving parts, the G8-2. Eugene M. Gluhareff was too far ahead of his time, however; his jet-powered Rogallo Wing, called the Porta-Jet, never caught on. He did, however, adapt it to a back-pack helicopter, the precursor of an unusual rotorcraft introduced at Oshkosh in 1978—the Eagle tip-mounted helicopter, which carries a Gluhareff jet at the tips of both rotor blades.

Perhaps the most popular engine for hang gliders to appear has been the Soarmaster, a 10-horsepower, 30-pound powerplant with a long shaft that extends backward to the rear of the kite. It has been successfully adapted to such gliders as the Electra Flyer Cirrus 5A, 5B, 3, and 2; the Olympus 160 and 180; the Delta Wing Phoenix 6B, 6C, and 8; the Eipper Cumulus 10; the Seagull Seahawk 170, 190, 10M, and 10.5M; the Sky Sports Sirocco: the Sensor; the UP Firefly, and the Wills SST, Alpha, and XC.

While a few fatalities have occurred with PHG's, another danger has been foot laceration, when pilots flying in prone position have made contact with the propeller blades whirring behind them. Soarmaster accordingly redesigned their unit, making the drive shaft 6 inches longer.

In 1978, the world's first powered hang glider meet was held at the little sport-flying airport of Perris Valley, in Southern California, where more than a score of PHG's showed up. Organized by Joe Diamond, of Costa Mesa, California, it was a real milestone in aviation, and drew such thoughtful observers as T. Claude Ryan, the man who built Lindbergh's Spirit of St. Louis ocean-hopper; Ladislao Pazmany, veteran homebuilt designer; and Bruce Carmichael, aeronautical engineer and historian of the ultralight movement. One Easy Riser pilot at the Perris Valley meet was Trip Mellinger, who the year before had flown his PHG across 26-mile-wide Cataline Channel.

There is no room here to cover all the fine new ultralight PHG's flying today (we're saving that for a whole new book to be published

Ken Striplin's Foot Launched Air Cycle powered hang glider.

soon by TAB!) but let's tip our hat to an old friend, Volmer Jensen, designer of the fine little VJ-22 Sportman amphibian homebuilt, who hung an engine on his rigid-wing hang glider, the VJ-23 Sun Fun, and called it the VJ-23E. On May 9, 1978, a British birdman named David Cook flew a VJ-23E across the English Channel from England to France, reversing the 1909 accomplishment of Louis Bleriot in his Channel Hopper. Cook landed in five steps, just 1 hour 15 minutes after leaving England, close to where three German social workers on holiday were having a picnic. When he told them where he had come from, Cook recalls, "they shook their heads and said 'All zer English are crazy!' "

As the hundreds of PHG's become thousands, a whole new generation of youths are logging flight time and building a pilot pool for the military to draw from should a national emergency occur. Perhaps by the time you read this the FAA will have decided how to categorize them, but already in Australia they're called Mininum Aircraft if they weigh 400 pounds or less, and in England, if the all up weight (AUW) is 1000Kg max, they're Featherweights. Most average from 70-120Kg, and are licensed and controlled by England's C.A.A.

Cream of the PHG's today would appear to be the Mitchell Wing, designed by Don Mitchell. The design is straightforward, with a 12-degree wing sweepback, the center section flat and outboard panels with 6-degree dihedral each. The rigid wing tapers in chord

John Chotia's Weedhopper powered hang glider may be the world's smallest airplane.

from five feet at the root to two feet at the tips. Span is 34 feet, airfoil section the NACA 23015. Standard controls are used, with the ailerons also serving as elevators or as stabilators used together. Drag rudders at the wingtips may be used independently or together as air brakes.

Mitchell Wing comes in kit form either as a foot-launched craft or with tricycle gear, and normally uses the McCulloch 101 engine. Manufacturer is James M. Meade's M Company of Porterville, in California's San Joaquin Valley. On April 14, 1979, at a Porterville PHG meet, Dick Clawson of Visalia, California, took off, climbed out of sight, and three hours later reappeared with a new altitude record. A sealed barograph showed he'd climbed 11,700 feet AGL (12,140 MSL), for an unofficial world record.

There was considerable excitement that day, after Clawson and another Mitchell wing pilot vanished into the haze nearly two miles above Porterville. Some one cried: "Look! He's falling!" Sure enough, two white objects were seen spinning earthward, tumbling over and over. An ambulance was summoned, and everyone prayed. Then, as the objects fell lower, they were seen to be merely pieces of paper that had been carried aloft on a dust devil. The drinks were on the house, along with a pot of homemade chili beans.

Chapter 16
Composite Materials

For years, the dream of homebuilders and designers of ultralight aircraft has been a light, strong, easily-fabricated material with the strength-to-weight proportions of a pelican's hollow wing bone. You didn't know that bird bones are hollow? And that with their tough, resilient feathers are stronger and lighter than any wing structure built by man? Well, almost—Dr. Paul MacCready's graceful Man-Powered Aircraft, the Gossamer Albatross Channel Hopper, has an amazing wingspan of 96 feet (more than a DC-9) and yet weighs only 55 pounds! With pilot aboard its wing loading is under .5 lb/sq. ft., comparable to that of nature's wandering albatross, largest of all living seabirds, with a wingspan of 11.5 feet. It can soar for endless hours without a single wing flap, deriving energy from the winds blowing over rolling waves, just as condors soar along mountain slopes.

Much of the success of the Gossamer Albatross lies in the use of advanced technology materials undreamed of in the time of the Wright Brothers. The MPA's wings are covered with Mylar polyester film, a tough, versatile plastic introduced to the world in 1952 by The Du Pont Company. Mylar's strength does not deteriorate with age, moisture, or temperature extremes. It does not yellow or become brittle and has high tensile, tear, and impact strength. Mylar is chemically similar to Dacron polyester fiber, another Du Pont product, but is produced in a continuous sheet form, while Dacron is spun as a fiber. Mylar comes in 24 different types of plain and coated film, with several different thicknesses and coatings for each type, so you have a total of 245 different products to choose from. (MacCready used the half-mil (0.0005″ thick) tensilized film normally used in computer tapes, video cassettes, etc.). This material is made in a way that minimizes stretch.

A still newer Du Pont product, Kevlar, commercially introduced in 1971, was used to fabricate Gossamer Albatross' stabilizing cords because of its low-stretch, light-weight, and high-strength characteristics. Kevlar is an aramid fiber with a unique combination of high strength, toughness, and stiffness, properties that are maintained at temperatures well above 300°F.

These aromatic polyamides are significantly different in both chemical and physical structure from other polyamides such as nylon. Woven into fabric, Kevlar is used in flak vests and other soft body armor and safety gloves because of their inherent cut- and heat-resistant properties. You'll also find Kevlar used in reforcement of radial tires, hoses, and V-belts.

Let's stop a moment and ask, just what is a composite structure we hear so much about in the homebuilt movement? In its broadest sense, it could be a homebuilt of welded-steel fuselage joined to a sheet-metal tailcone, and fabric-covered. In the current sense, however, a composite structure is one made of fiberglass and epoxy over a foam or wood core. European sailplane makers were pioneers in the use of the fiberglass-foam manufacturing technique.

The styrene foam used most widely in homebuilts is the type that comes in slabs for construction and insulation of homes, and is used as a filler to give contour to a basic wooden core, such as that as the Rand-Robinson KR-1 and KR-2, with a density of 3 lb. sq. ft. and bubbles small enough for good bonding. It can easily be cut, sanded, glued, and hot-wire sawn.

Dynel fabric, manufactured by Union Carbide Corp., is a popular and inexpensive material for skinning a wing or a fuselage styrofoam substructure, used in combination with epoxy resin. Epoxy has good strength and stickiness and low shrinkage during cure. Dynel is lighter and stronger, pound for pound, than fiberglass and can be stretched more easily without tearing.

Fiberglass, introduced prior to World War II, possesses high strength, dimensional stability, and is resistant to flame. Glass fabrics come with different weaves that determine the drape characteristics and warp and fill pattern.

Presence of moisture will chemically degrade the interface between the fiber surface and the resin used, resulting in a dramatic reduction in strength, hence a number of special finishes have been developed for such resin systems as melamine, silicone, polyester, phenolic, epoxy, and polimide. Be sure that the finish being used is compatible with the resin, to avoid degradation due to dampness.

Let's turn back a moment for a historical look at the nature and use of plastics in aircraft manufacture, that really began with the Duramold process used by Howard Hughes in his giant flying boat, the HK-1 "Spruce Goose" in World War II. A little later, initial research in fiberglass aircraft construction was launched in 1955 by August Bellanca, son of Giuseppe Bellanca, who founded his own research firm, Bellanca Aircraft Engineering Inc., on Long Island, N.Y., in 1963.

The Bellancas, father and son, dreamed of building a plastic plane for four, with retractable gear and a 400-horsepower Lycoming engine, that could cruise at 260 mph with a range of 1100 milles. Their pioneer work was taken over by a Texas firm, Payne Engineering, which built a 6-place prototype called Skyrocket II, that set some FAI speed records, including the 500-kilometer closed course event in which it was clocked at 327.7 mph at 30,000 feet.

Skyrocket II actually used an airframe structure of fiberglass, epoxy impregnated, over an aluminum sandwich core.

Another interesting plastic plane project was launched in Midland, Texas—the Windecker X-7 Eagle I, a four-place low-winger powered with a 290-horsepower Lycoming I0-540G engine. Its creator was Dr. Leo Windecker a dentist, who with his wife, also a DDS, practiced their profession at the Dow Chemical Company clinic in Freeport, Texas, where they began experimenting with some new Dow plastics for possible use in making dental bridgework. Dr. Windecker learned to fly and soon was experimenting at Hondo, Texas, at a subsidiary Dow firm, Windecker Research Inc., on plastic aircraft materials.

Recalling the amazing strength-to-weight ratio of bones, he developed a hard, cellular material that could be machined into airfoil sections with holes to slip over spars. Plastic sheet wrapped around the plastic foam core. The wing was test-flown on an Ercoupe and worked well. For the fuselage he went to thermal setting epoxy adhesives laid up with fiberglass cloth, which he called Fibaloy. Fuselage sides were laid up in plastic molds. Eagle I first flew on October 7, 1967, and proved to be a responsive, stable, quiet aircraft, but a subsequent accident halted the program.

By the mid-1970s, plastics had come into widespread use in American lifestyles, and close to two billion pounds of reinforced plastics had found their way to the marketplace. It soon appeared that still more revolutionary materials would come into use in the 1980s—advanced composites possessing characteristics of low weight, high strength, and stiffness, in many respects superior to metals for sophisticated aircraft use. There appeared a whole new class of reinforcement fibers, such as tungsten substrate, boron, rayon, graphite, and carbon filaments. Although some half billion dollars had been expended to develop this advanced technology, homebuilders were slow to adopt it due to its high cost. One who did is Burt Rutan, designer of the popular VariViggen, VariEze, Quickie, and Defiant, who planned to use the advanced materials in the latter production aircraft.

Today, the most common fiber material used in homebuilt aircraft is, of course, fiberglass, and has been used for years in such non-structural items as wheel pants, cowlings, and wingtips. Now, the entire structure of the Rutan VariEze is a composite of high-strength, primarily unidirectional fiberglass, with rigid foam as core material.

All-glass aircraft structures have been certified in the United States since 1964, in the 301 Libelle sailplane. The VariEze glass structure is quite similar to that of many glass sailplanes: epoxy resin (Shell epon 815) and hand-laidup glass cloth with room-temperature cure. However, Rutan points out, the extensive use of composite construction (two high-strength layers separated by a low-density core) and the time saving methods of fabrication employed, are entirely new.

Glass/foam/glass composite advantages include:

- Higher strength-to-weight ratio when properly applied.
- Quiet—no "oil canning" of any surface; dampens engine noise.
- Provides insulation—even the pilot's feet stay warm in cold weather.
- Higher fatigue margin—no concentration of loads, no sonic panel vibration.
- Contour formation simplified—compound shapes easier to form.
- Contour is maintained under flight load.
- Easy to build—one layup replaces many parts and fasteners.

Glass construction, Rutan points out, paces the construction time, since once a specific layup is started and the epoxy is mixed, that job must be completed within the pot time. The VariEze wings have only two layups (or two epoxy-mix periods), thus it takes only two days to build a wing—and it is not possible to spend three days at the job, therefore! The wings are designed to stiffness criteria and

Pair of Rutan Vari Ezes in flight over Mojave Desert.

for ground-handling loads, rather than to ULF (Ultimate Load Factor) flight loads. Thus, when enough glass material is applied to stiffen the wing to eliminate flutter, the ultimate G capability is over 16 G's. The result is a much greater strength and fatigue margin than that of a conventional wood or metal structure.

VariEze fuselage construction also is "Vari Eze"— consisting of glassing one side of large foam sheets on a horizontal surface, sawing out major parts (bulkheads, sides, and bottom), assembling the parts, rounding the outer edges, and glassing the outside to complete the composite. Use of special weave glass cloth and variable density resin results in layups that flow down and squeegy out with ease.

The first resins to come into popular use with homebuilders were the polyesters, frequently called boat resins because of their widespread use in constructing boat hulls. They are bluish in color, or may be clear, depending on the additives, and the pot life is short—they will harden within half an hour after adding the hardening catalyst. Polyester resin should not be used over any foam plastic except polyurethane. However, it is in wide use with either fiberglass or Dynel cloth for laminated layups and for making nonstructural parts such as wheel pants or cowlings. When working with a female mold, homebuilders usually apply a thick spreading of pigmented Gel-Coat polyester resin prior to the laminating process, producing a smooth surface on the finished product.

Observed under a microscope, raw polyester resin looks like a pile of chain links (polyester molecules) and individual links (monomers). When a catalyst is added, the monomer links join and crosslink between the polyester molecules, in a process called polymerization, resulting in the liquid resins gelling into a solid, rigid plastic.

Epoxy resins, due to their greater strength, generally have replaced polyesters in layup and sprayup applications. Epoxies are more hazardous to work with than the polyester resins, because they contain chemicals which are powerful skin irritants and sensitizers. Basic epoxy is too thick to use as a laminating resin, hence additives are used to thin it. The hardener-to-resin ratio by weight may vary from 1:1 to 1:10.

Strict humidity control is necessary to get a good epoxy mix, but most epoxies have a pot life of one hour or more and will harden in one day at 72°F. Rutan recommends working in temperatures between 65°F and 85°F, perhaps higher, but never lower than that range. Among other tips, he suggests installing a 25-watt light bulb near the epoxy so that it will wet the cloth faster. When laying out fiberglass for work, spread it on a clean board and roll it out. A kitchen table is fine but keep your work area free from moisture and dirt contamination.

Fiberglass for non-structural applications comes in four types—chopped and formed into a sheet called fiberglass mat; woven into long filaments; interwoven into cloth: and cloth impregnated already by the maker. Mat is used for moldmaking, being quite heavy when saturated with resin. Fiberglass cloth of 1.8 to 7 ounce weight is most widely used in homebuilding, providing the best strength-to-weight ratio.

Hollmann Sportster is only two-place homebuilt gyrocopter.

Says Rutan: "The basic material in your VariEze is glass cloth, which is available in hundreds of different weights, weaves, strengths, and working properties. Very few of the commercial glass cloths are compatible with aircraft requirements for high strength and light weight. Even fewer are suitable for the hand layup techniques we have developed for the homebuilder.

"The glass cloth in the VariEze carries primary loads, and its correct application is vital. Two types of glass cloth are used—a bi-directional cloth (RA5277BID) and a uni-directional cloth (RA5177UND). BID cloth has half the fibers woven parallel to the selvage edge of the cloth and the other half at right angles to the selvage, giving the cloth the same strength in both directions.

"UND cloth has 95% of the glass fibers woven parallel to the selvage, giving exceptional strength in that direction and very little at right angles to it. BID is generally used for pieces which are cut at a 45-degree angle to the selvage, a bias cut, which enables you to lay a BID into contours with very little effort. UND is used in areas where the primary loads are in one direction and maximum efficiency is required, such as the wing skins and spar caps. Multiple layers of cloth are laminated together to form the aircraft structure. Each layer of cloth is called a ply."

There is no universal epoxy system, Rutan says. Each one has its own specific purpose, which may be best for one application but worst for another. The RAE-type epoxy systems recommended for use in building a VariEze are tailored for the best combination of workability and strength. These epoxies are not similar to the common types normally marketed for fiberglass lamination. RAE is a specially modified type that offers a tenfold increase in fatigue life. Three different systems are used in VariEze construction for three different types of work— a slow-curing system, a fast-curing system, and a 5-minute curing system. The latter is used the way clecos are used in sheet metal construction, or clamps are used in woodwork, for temporary positioning, as well as in some areas where high strength is not required but where fast curing will aid assembly.

As an epoxy system cures it generates heat, and, in some areas, the heat buildup of a medium or fast curing epoxy is definitely unacceptable. Where this is a potential problem, a slow-curing system is used. Slow cure is always used with styrofoam, where heat can melt away the foam and ruin the joint. In other areas, where heat buildup is not a problem and a faster cure is desirable, the fast cure system is employed. Both fast and slow cure epoxies will cure to a firm structure overnight at room temperature (70°F to 80°F).

Rutan Defiant push-pull twin is latest glass plane to fly.

The working and strength characteristics of an epoxy system are dependent upon the resin, the hardener, and the amount of each in a given mixture. Epoxy systems are engineered for a specific ratio to hardener. It's essential that you get the proper mixture, and an accurate balance or ratio pump must be used. Ratio pumps are available commercially for around $130, or you can ratio the epoxy with a homemade scale. In operation, a five-step procedure must be followed precisely each time a batch of epoxy is mixed:

- Place both empty paper cups on balance as shown.
- Adjust ballast weight to level mark.
- Fill resin cup with desired amount of resin, from 1 to 7 ounces.
- Add hardener to hardener cup to balance scale on the level mark, then add a few more drops of hardener, since it is impossible to pour all the hardener from the cup.
- Pour hardener into resin cup and mix.

Epoxy resin and hardener are mixed in small batches, usually 6 ounces or less, even in the largest layup. This is because, in mixing larger batches, as the hardening reaction progresses, heat is generated and this speeds up the reaction, which produces still more heat, ending up in an undesirably fast reaction called an *exotherm.*

An exotherm will cause the cup of epoxy to overheat and thicken rapidly, and should this occur, throw it away and start over.

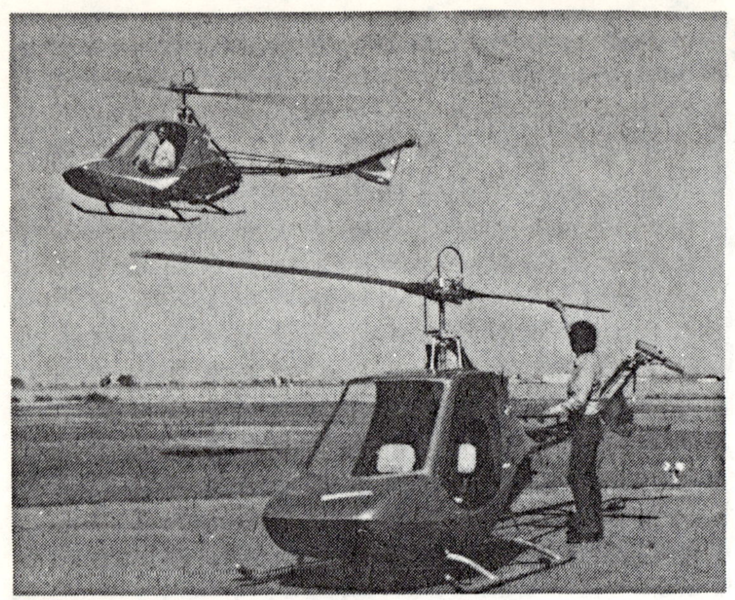

Scorpion two-place homebuilt helicopters are popular.

Use unwaxed paper cups for mixing and rationing epoxy and hardener. VariEze distributors carry 8-oz cups for the resin, and 3-oz unwaxed bathroom paper cups are available at your favorite supermart. (Waxed cups will contaminate the epoxy and should never be used.)

What direction the technology of modern composite materials will go in the future is anybody's guess, but there is little question that still lighter, stronger structures will become available to the dedicated homebuilder seeking the ultimate in strength-to-weight ratios and ease of handling.

Perhaps the ultimate material already *has* been discovered—in 1893, at an International Conference on Aerial Navigation held in Chicago during the World's Fair that year, one enthusiast claimed the secret of bird flight lay in goose feathers. "Take a six-pound goose and strip away six ounces of feathers and he cannot fly at all," he pointed out. And as early as 1776, Benjamin Franklin proposed harnessing a flock of geese to a framework to transport a man through the air. Just maybe—?

Appendix

The following data are presented to give the reader further details of homebuilt aircraft described in this book, including specifications, performance, and addresses, where available.

ACEY DEUCY

Engine: Continental A65 or larger. Wing Span 32′ 6″. Wing area 155 sq.ft. Length 20′ 9″. Height 6′ 9″. Empty weight 750 lb. Gross weight 1275 lb. Vmax 98 mph. Max cruise 83 mph. ROC 450 fpm. Range 250 miles. Contact: John Powell, 4 Donald Drive, R.I. 02840.

STOLP SA-700 ACRODUSTER 1

Engine: Lycoming I0-360. Wing span 19′. Wing area 105 sq.ft. Length 15′ 9″. Height 6′ 3″. Empty weight 740 lb. Gross weight 1190 lb. Vmax 180 mph. Max cruise 165 mph. ROC 3000 fpm. Range 360 miles. Contact: Stolp Starduster Corp., 4301 Twining Riverside CA 92509.

STOLP SA-750 ACRODUSTER TOO

Engine: Lycoming I0-360 (200 hp). Wing span 21′ 5″. Wing area 130 sq. ft. Length 18′ 6″. Height 6′ 10″. Empty weight 1000 lb. Gross weight 1541 lb. Vmax 200 mph. Max cruise 160 mph. ROC 2300 fpm. Range 640 miles. Contact: Stolp Starduster Corp.

STOLP STARDUSTER TOO

Engine: Lycoming I0-360 (180 hp). Wing span 24′. Wing area 165 sq ft. Length 20′ 5″. Height 7′ 4″. Empty weight 1000 lb. Gross weight 1704 lb. Vmax 148 mph. Max cruise 134 mph. ROC 1500 fpm. Range 600 miles. Contact: Stolp Starduster Corp.

ANDERSON KINGFISHER

Engine: Continental 0-200. Wing span: 37′ 1″. Wing area 185 sq ft. Length 23′ 7″.Height 8′ 4″. Empty weight 1092 lb. Gross weight 1600 lb. Vmax 113 mph. Max cruise 90 mph. ROC 650 fpm. Range 330 miles. Contact: Earl W. Anderson, 4708 Arlington Rd., Palmetto, FL 33561.

BABY LAKES

Engine: Continental A-80. Wing span 16′ 8″. Wing area 86 sq. ft. Length 13′ 9″. Height 4′ 6″. Empty weight 475 lb. Gross weight 850 lb. Vmax 135 mph. Max cruise 118 mph. ROC 2000 fpm. Range 250 miles. Contact: Barney Oldfield Aircraft Co., P.O. Box 5974, Cleveland OH 44101.

BENSEN B-8M GYROCOPTER

Engine: 72 thru 90 hp. Main rotor span 21′ 8″. Length 11′ 3″. Height 6′ 9″. Empty weight 247 lb. Gross weight 500 lb. Vmax 85 mph. Cruise 60 mph. ROC 600-1000 fpm. Range 100 miles. Contact: Bensen Aircraft Co., P.O. Box 31047, Raleigh N.C. 27611.

BREEZY RLU-1

Engine: Continental C-90. Wing span 33′. Wing area 165 sq. ft. Length 22′ 6″. Height 8′ 6″. Empty weight 700 lb. Gross weight 1200 lb. Vmax 105 mph. Cruise 75 mph. ROC 500 fpm. Range 250 miles. Contact: Charles B. Roloff, 8025 W. 90th St., Hickory Hills IL 60457.

CHRISTEN EAGLE I

Engine: Lycoming AEIO-540 (260 hp). Wing span 19′ 11″. Wing area 124 sq. ft. Length 17′ 11″. Height 6′ 6″. Empty weight 997 lb. Gross weight 1478 lb. Vmax 197 mph. Max Cruise 171 mph. ROC 2640 fpm. Range: 350 miles std. Contact: Christen Industries, Inc. 1048 Santa Ana Valley Rd., Hollister CA 95023.

CHRISTEN EAGLE II

Engine: Lycoming AEIO-360-AID (200 hp). Wing span 19′ 11″. Wing area 125 sq. ft. Length 17′11″. Height 6′ 6″. Empty weight 1025 lb. Gross weight 1578 lb. Vmax 184 mph. Max cruise 165 mph. ROC 2100 fpm. Range 380 miles. Contact: Christen Industries, Inc.

CORBEN BABY ACE D

Engine: Continental C-85. Wing span 26′ 5″. Wing area 112/3 sq. ft. Length 17′ 8-3/4″. Height 6′ 7-3/4″. Empty weight 575 lb. Gross weight 900 lb. Vmax 130 mph. Max cruise 90 mph. ROC 1200 fpm. Range 350 miles. Contact: Ace Aircraft Mfg. Co., 106 Arthur Rd., Asheville N.C. 28806.

EAA ACRO SPORT

Engine: 100 hp thru 180 hp. Wing span 19′ 7″. Wing area 116 sq ft. Length 17′ 6″. Height 6′. Empty weight 733 to 926 lb. Gross weight 1350 lb. Vmax 180 mph. Max cruise 130 mph. ROC 3000 fpm. Range 300 miles. Contact: Acro Sport, Inc., PO Box 462, Hales Corners WI 53130.

EAA BIPLANE

Engine 85 thru 125 hp. Wing span 20′. Wing area 120 sq. ft. Length 17′. Height 6′. Empty weight 640 lb. Gross weight 1023 lb. Vmax 125 mph. Max cruise 110 mph. ROC 1000 fpm. Range 200 miles. Contact: EAA Air Museum Foundation, P.O. Box 469, Hales Corners WI 53130.

EVANS VP-1

Engine: VW 1600 cc thru 2100 cc. Wing span 24′. Wing area 100 sq. ft. Length 18′. Height 5′. Empty weight 440 lb. Gross weight 750 lb. Vmax 95 mph. Max cruise 75 mph. ROC 600 fpm. Range 200 miles. Contact: Evans Aircraft, Box 744, La Jolla CA 92037.

EVANS VP-2

Engine: VW 1800cc & 2100 cc. Wing span 27′. Wing area 130 sq ft. Length 19′. Height 5′6″. Empty weight 640 lb. Gross weight 1040 lb. Vmax 100 mph. Max cruise 75 mph. ROC 500 fpm. Range 300 miles. Contact: Evans Aircraft.

FIREFLY

Engine: Rolls Royce C-85. Wing span 19′ 2″. Wing area 80 sq. ft. Length 18′ 11″. Height 7′. Empty weight 835 lb. Gross weight 1300 lb. Vmax 160 mph. Max cruise 152 mph. ROC 800 fpm. Range 450 miles. Contact: Bill Statler Sr. 9300 Encino Ave., Northridge CA 91324.

FLAC

Engine: CGS Power Hawk. Wing Span 32′. Wing area 155 sq. ft. Empty weight 156 lb. Gross weight 381 lb. Vmax 60 mph. Max cruise 55 mph. Best L/D speed 36 mph. Minimum sink 150 fpm. Contact: Striplin Aircraft Corp., P.O. Box 2001, Lancaster CA 93534.

FLAGLOR SCOOTER

Engine: VW 1500 cc, 40 hp. Wing span 28′. Wing area 115 sq. ft. Length 15′ 8″. Height 7′. Empty weight 390 lb. Gross weight 650 lb. Vmax 90 mph. Max cruise 80 mph. ROC 600 fpm. Range 175 miles. Contact: Ace Aircraft Mfg. Co., 106 Arthur Road, Asheville, N.C. 28806.

FLY BABY

Engine: 65 thru 100 hp. Wing span 22′. Wing area 150 sq. ft. Length 18′ 10-1/2″. Height 6′ 11″. Empty weight 650 lb. Gross weight 950. Vmax 115 mph. Max cruise 85 mph. ROC 800 fpm. Range 330 miles. Contact: Peter M. Bowers, 10458 16th Ave. S., Seattle WA 98168.

GOSSAMER CONDOR

Engine: Man-powered. Wing Span 96′. Wing area 800 sq. ft. Length 30′. Height 19′ 6″. Empty weight 70 lb. Gross weight 220 lb. Vmax 12 mph. Cruise 10 mph. Climb NA. Range NA. Contact: Paul MacCready, 145 Vista Ave., Pasadena CA 91107.

HATZ CB-1 BIPLANE

Engine: Lycoming 0-320 (150 hp). Wing span 26′. Wing area 190 sq. ft. Length 18′ 6″. Height 7′ 10″. Empty weight 966 lb. Gross weight 1600 lb. Vmax NA. Max cruise 100 mph. ROC 1200 fpm. Range 270 miles. Contact: Dudley R. Kelly, Route 4, Versilles, KY. 40383.

HOLLMANN HA-2M SPORTSTER

Power: 130 thru 150 hp. Rotor blade span 28′. Length 12′. Height 7′ 8″. Empty weight 620 lb. Gross weight 1050 lb. Vmax 95 mph. Max cruise 50 mph. ROC 500 fpm. Range 90 miles. Contact: Hollmann Aircraft, 7917 Festival Ct., Cupertino CA 95014.

ISAACS FURY

Engine: 65 thru 125 hp. Wing span 21′. Wing area 124 sq. ft. Length 19′ 3″. Height 7′ 1″. Empty weight 720 lb. Gross weight 1000 lb. Vmax 115 mph. Max cruise 90 mph. ROC 1600 fpm. Range 150 miles. Contact: John O. Isaacs, 23 Linden Grove, Chandler's Ford, Hampshire, S05 ILE, England.

ISAACS SPITFIRE

Engine: 0-200 (100 hp). Wing span 22′ 1-1/2″. Wing area 87 sq. ft. Length 19′ 4″. Height 7′ 2″. Empty weight 805 lb. Gross weight 1100 lb. Vmax 150 mph. Max cruise 133 mph. ROC 1100 fpm. Range 200 miles. Contact: John O. Isaacs.

JURCA GNATSUM

Engine: 125 thru 200 hp. Wing span 25′ 10″. Wing area 110 sq. ft. Length 22′ 4″. Height NA. Empty weight 1175 lb. Gross weight 2000 lb. Vmax 225 mph. Max cruise 175 mph. ROC 1500 fpm. Range 570 miles. Contact: Betty Heit, Jurca Plans, 6732-18 Maple Brook, Flint, MI 48507.

JURCA MJ-5 SIROCCO

Engine: Lycoming 0-235 (115 hp). Wing span 19′ 6″. Wing area 85 sq. ft. Length 18′ 6″. Height 8′. Empty weight 726 lb. Gross weight

1210 lb. Vmax 137 mph. Max cruise 124 mph. ROC 1150 fpm. Range 570 miles. Contact: Betty Heit, Jurca Plans.

JT-5 AUTOGYRO

Engine: VW 1700 cc. Rotor blade span 23′. Length 11′ 6″. Height 6′ 6″. Empty weight 378 lb. Gross weight 639 lb. Vmax 93 mph. Max cruise 81 mph. ROC 540 fpm. Range 186 miles. Contact: Jukka Tervanaki, Aidasmaentie 16-20E, 00650 Helsinki 65, Finland.

KR-1

Engine: VW 1800 cc. Wing span 17′. Wing area 62 sq. ft. Length 12′ 9″. Height 3′ 8″. Empty weight 370 lb. Gross weight 750 lb. Vmax 140 mph. Max cruise 130 mph. ROC 900 fpm. Range 748 miles. Contact: Rand Robinson Engineering, 5842 McFadden, Unit K, Huntington Beach CA 92649.

KR-2

Engine: Revmaster 2100 cc, turbocharged. Wing span 20′ 8″. Wing area 78 sq. ft. Length 14′ 6″. Height 3′ 8″. Empty weight 480 lb. Gross weight 900 lb. Vmax 200 mph. Max cruise 190 mph. ROC 1200 fpm. Range 1000 miles. Contact: Rand Robinson Engineering.

LARKIN SKYLARK

Engine: VW 1600 cc. Wing span 26′. Wing area 114 sq. ft. Length 19′ 6″. Height 6′ 2-1/2″. Empty weight 790 lb. Gross weight 1246 lb. Vmax 115 mph. Max cruise 105 mph. ROC 550 fpm. Range 525 miles. Contact: Larkin Aircraft Corp., P.O. Box 66899 Scotts Valley CA 95066.

MARQUART MA-5 CHARGER

Engine: 100 thru 180 hp. Wing span 24′. Wing area 170 sq. ft. Length 19′ 6″. Height 7′ 6″. Empty weight 1000 lb. Gross weight 1550 lb. Vmax 125 mph. Max cruise 115 mph. ROC 1200 fpm. Range 450 + miles. Contact: Ed Marquart, 3663 Grandview, Riverside CA 92509.

MIDGET MUSTANG

Engine: 85 thru 150. Wing span: 18′ 6″. Wing area 68 sq. ft. Length 16′ 5″. Height 4′ 6″. Empty weight 560 lb. Gross weight 960 lb. Vmax 225 mph. Max cruise 215 mph. ROC 2200 fpm. Range 375 miles. Contact: Bushby Aircraft, Inc., Rt. 1 Box 13, Minooka IL 60447.

MINI COUPE

Engine: VW 1600 cc. Wing span 24′. Wing area 83-1/2 sq. ft. Length 16′ 4″. Height 5′ 11″. Empty weight 497 lb. Gross weight 850 lb. Vmax 105 mph. Max cruise 90 mph. ROC 750 fpm. Range 300 miles. Contact: Chris Tena Aircraft. P.O. Box 1, Hillsboro, OR 97123.

MITCHELL WING

Engine: McCulloch 101. Wing span 34′. Wing area 136 sq. ft. Airfoil NACA 230-15. Aspect ratio 8:1. Max design load 1800 lb. Max cruise 55 mph. ROC 200+ fpm. Max L/D 36 mph. Glide ratio 16:1. Weight 140 lb. Contact: M Company, 1900 So. Newcomb, Porterville CA 93257.

MOODY EASY RISER

Engine: McCulloch 101. Wing span 30′. Wing area 170 sq. ft. Airfoil LM 7610. Aspect ratio 8.8:1. Speed range 16-45 mph. Glide ratio 10:1. Weight 50 lb. ULF 7.5G. Contact: UFM Ultralight Flying Machines, P.O. Box 59, Cupertino CA 95014.

NEOVA II

Engine: VW 1600 cc. Length 14′. Width 7′. Height 55″. Empty weight 750 lb. Gross weight 1110 lb. Cruise (water) 32 mph; (ice) 45 mph. Contact: Neoteric-USA Inc., Fort Harrison Industrial Park, Terre Haute In 47804.

OSPREY II

Engine: Lycoming 0-320 (150 hp). Wing span 26′. Wing area 130 sq. ft. Length 20′ 6″. Height 6′. Empty weight 970 lb. Gross weight 1560 lb. Vmax 150 mph. Max cruise 130 mph. ROC 1200 fpm. Range 300 miles. Contact: Osprey Aircraft, 3741 El Rincon Way, Sacramento CA 95825.

PIETENPOL AIR CAMPER

Engine: Continental A-65. Wing span 29′. Wing area 145 sq. ft. Length 18′ 9″. Height 6′ 6″. Empty weight 622 lb. Gross weight 1200 lb. Vmax 100 mph. Max cruise 80 mph. ROC 500 fpm. Range 200 miles. Contact: B. H. Pietenpol, Spring Valley MN 55975.

PITTS SPECIAL

Engine: Lycoming IO-360. Wing span 17′ 5″. Wing area 98 sq. ft. Length 15′ 6″. Height 6′. Empty weight 720 lb. Gross weight 1050 lb. Vmax 203 mph. Max cruise 154 mph. ROC 2640 fpm. Range 300 miles. Contact: Pitts Aerobatics, P.O. Box 547, Afton WY 83110.

POBER PIXIE

Engine: 50 thru 75 hp. Wing span 30′. Wing area 134-1/4 sq. ft. Length 17′ 3″. Height 6′ 2″. Empty weight 537 lb. Gross weight 950 lb. Vmax 130 mph. Max cruise 85 mph. ROC 700 fpm. Range 300 miles. Contact: Acro Sport Inc., PO Box 462, Hales Corners. WI 53130.

REDFERN FOKKER DR-1

Engine: 115 thru 165 hp. Wing span 23′ 7″. Wing area 202 sq. ft. Length 19′. Height 9′ 8″. Empty weight 948 lb. Gross weight 1378 lb. Vmax 120 mph. Max cruise 100 mph. ROC 2000 fpm. Range 300 miles. Contact: W. W. Redfern, Rt. 1, Athol ID 83801.

REDFERN NIEUPORT 17

Engine: 145 hp Warner. Wing span 26′ 11″. Wing area 165.45 sq. ft. Length 18′ 10″. Height 7′ 8″. Empty weight 1004 lb. Gross weight 1390 lb. Vmax 160 mph. Max cruise 115 mph. ROC 1000 fpm. Range 300 miles. Contact: W. W. Redfern, Rt. 1, Athol, ID 83801.

RUTAN VARI EZE

Engine: Continental 0-200. Wing span 22′ 4″. Wing area 53.6 sq. ft. Length 12′ 5″. Height 6′ 1″. Empty weight 535 lb. Gross weight 1050 lb. Vmax 211 mph. Max Cruise 201 mph. ROC 1700 fpm. Range 1100 miles. Contact: Rutan Aircraft Factory, P.O. Box 656, Mojave CA 93501.

RUTAN VARI VIGGEN

Engine: 150 thru 180 hp. Wing span 23′ 9″. Wing area 125 sq. ft. Length 20′. Height 6′ 1″. Empty weight 960 lb. Gross weight 1860 lb. Vmax 175 mph. Max cruise 164 mph. ROC 1300 fpm. Range 780 miles. Contact: Rutan Aircraft Factory.

VAN'S RV-3

Engine: 85 thru 125 hp. Wing span 19′ 11″. Wing area 90 sq ft. Length 19′. Height 5′ 3″. Empty weight 695 lb. Gross weight 1050 lb. Vmax 195 mph. Max cruise 185 mph. ROC 1900 fpm. Range 600 miles. Contact: Van's Aircraft, 22730 S.W. Francis, Beaverton OR 97005.

SCORPION 133

Engine: RW 133. Main rotor span 25′. Length 20′ 6″. Height 7′ 3″. Seats - 2. Empty weight 790 lb. Gross weight 1220 lb. Vmax 90 mph. Max cruise 75 mph. ROC 1000 fpm. Range 150 miles. Contact: RotorWay Aircraft, Inc., 14805 Interstate 10, Tempe AZ 85284.

SINDLINGER HAWKER HURRICANE

Engine: 150 thru 180 hp. Wing span 25′. Wing area 100 sq. ft. Length 19′ 8″. Height 5′ 10″. Empty weight 1000 lb. Gross weight 1375 lb. Vmax 198 mph. Max cruise 175 mph. ROC 1200 fpm. Range: 650 miles. Contact: Sindlinger Aircraft, 5923 9th St., NW, Puyallup, WA 98371.

SPENCER AMPHIBIAN AIR CAR

Engine: Continental Tiara (285 hp). Wing span 37′ 4″. Wing area 184 sq. ft. Length 26′ 5″. Height 9′ 6″. Empty weight 2190 lb. Gross weight 3200 lb. Vmax 147 mph. Max cruise 140 mph. ROC 1000 fpm. Range: 800 miles. Contact: P. H. Spencer, 8725 Oland Ave., Sun Valley CA 91352.

STEEN SKYBOLT

Engine: Lycoming HO-360 (180 hp). Wing span 24′ (upper). Wing area 152.7 sq. ft. Length 19′. Height 7′. Empty weight 1080 lb. Gross weight 1680 lb. Vmax 145 mph. Max cruise 130 mph. ROC 2500 fpm. Range 450 miles. Contact: LaMar Steen, 3218 Cherry St., Denver CO 80222.

STEWART FOO FIGHTER

Engine: Franklin Sport 4 (125 hp). Wing span 20′ 8″. Wing area 140 sq. ft. Length 18′ 9″. Height 7′. Empty weight 725 lb. Gross weight 1100 lb. Vmax 125 mph. Max cruise 115 mph. ROC 1200 fpm. Range 250 miles. Contact: Don Stewart, 11420 Route 165, Salem OH 44460.

TAYLOR COOT

Engine: 120 thru 220 hp. Wing span 36′. Wing area 180 sq. ft. Length 22′. Height 8′. Empty weight 1250 lb. Gross weight 1950 lb. Vmax 140 mph. Max cruise 120 mph. ROC 1250 fpm. Range 500+ miles. Contact: M. B. (Molt) Taylor, Box 1171, Longview WA 98632.

TEENIE TWO

Engine: 1500 cc VW. Wing span 18′. Wing area 60 sq. ft. Length 12′ 10″. Height NA. Empty weight 310 lb. Gross weight 590 lb. Vmax 120 mph. Max cruise 110 mph. ROC 800 fpm. Range 400 miles. Contact: Cal Parker, P.O. Box 625, Coolidge, AZ 85228.

THORP T-18

Engine: Lycoming IO-360. Wing span 20′ 10″. Wing area 86 sq. ft. Length 11′ 8″. Height 5′ 1″. Empty weight 900 lb. Gross weight 1500 lb. Vmax 209 mph. Max cruise 184 mph. ROC 1800 fpm. Range 580 miles. Contact: Thorp Engineering Co., Drawer T, Lockeford CA 95237.

VOLMER VJ-22 SPORTSMAN

Engine: Continental C-85. Wing span 36′ 6″. Wing area 175 sq. ft. Length 24′. Height 8″. Empty weight 1000 lb. Gross weight 1500 lb. Vmax 95 mph. Max cruise 85 mph. ROC 600 fpm. Range 300 miles. Contact: Volmer Aircraft, P.O. Box 5222, Glendale CA 91201.

WAG AERO CUBy

Engine: 65 thru 125 hp. Wing span 35′ 2-1/2″. Wing area 178.5 sq. ft. Length 22′ 2-3/4″. Height 6′ 8″. Empty weight 730 lb. Gross weight 1400 lb. Vmax 122 mph. Max cruise 94 mph. ROC 490 fpm. Range 825 miles (w/aux). Contact: Wag Aero, Inc., P.O. Box 181, Lyons, WI 53148.

WAG-A-BOND

Engine: 65 thru 100 hp. Wing span 29′ 3″. Wing area 147.5 sq. ft. Length 18.7′. Height 6′. Empty weight 640 lb. Gross weight 1250 lb. Vmax 105 mph. Max cruise 95 mph. ROC 625 fpm. Range 300 miles. Contact: Wag Aero, Inc.

WAR REPLICAS F/W 190

Engine: Continental 0-200. Wing span 20′. Wing area 70 sq ft. Length 16′ 7″. Height 5′. Empty weight 600 lb. Gross weight 900 lb. Vmax 195 mph. Max cruise 145 mph. ROC 1100 fpm. Range 325 miles. Contact: War Aircraft Replicas Inc., 348 S. 8th St., Santa Paula CA 93060.

WEEDHOPPER

Engine: Chotia 2-cycle (18 hp). Wing span 28′. Wing area 168 sq. ft. Empty weight 160 lb. Max load 220 lb. Vmax 50 mph. Max cruise 30 mph. ROC 600 fpm. Range 30 miles. Contact: Weedhopper of Utah, Inc., P.O. Box 2253, Ogden UT 84404.

WICHAWK

Engine: 125 thru 300 hp. Wing span 24′. Wing area 185 sq. ft. Length 19′ 3″. Height 7′ 2″. Empty weight 1280 lb. Gross weight 2400 lb. Vmax 145 mph. Max cruise 135 mph. ROC 2500 fpm. Range 500 miles. Contact: Javelin Aircraft Co. Inc., 1978 Easy St., Wichita KS 67230.

WITT'S VEE

Engine: VW 1600 cc. Wing span 17′ 6″. Wing area 77 sq. ft. Length 18′ 2″ Height 4′. Empty weight 430 lb. Gross weight 700 lb. Vmax 200 mph. Max cruise 150 mph. ROC 1000 fpm. Range 450 miles. Contact: Steve Wittman, Box 2762 Oshkosh WI 54901.

Index